THE ELECTRONIC DISTURBANCE

Autonomedia New Autonomy Series

Jim Fleming & Peter Lamborn Wilson, Editors

THE
ELECTRONIC
DISTURBANCE

CRITICAL ART ENSEMBLE

AUTONOMEDIA

Critical Art Ensemble would like to thank
Jim Fleming and all the members
of the Autonomedia Collective who helped
bring this project to fruition. We would
especially like to thank Steven Englander,
whose editorial assistance was invaluable.

Autonomedia
POB 568 Williamsburgh Station
Brooklyn, NY 11211-0568 USA

718-387-6471

Printed in the United States of America

CONTENTS

1

Introduction

The
Virtual
Condition

The Virtual Condition is a Tele-gamble That Always Spins Off

The rules of cultural and political resistance have dramatically changed. The revolution in technology brought about by the rapid development of the computer and video has created a new geography of power relations in the first world that could only be imagined as little as twenty years ago: people are reduced to data, surveillance occurs on a global scale, minds are melded to screenal reality, and an authoritarian power emerges that thrives on absence. The new geography is a virtual geography, and the core of political and cultural resistance must assert itself in this electronic space.

The West has been preparing for this moment for 2,500 years. There has always been an idea of the virtual, whether it was grounded in mysticism, abstract analytical thinking, or romantic fantasy. All of these approaches have shaped and manipulated invisible worlds accessible only through the imagination, and in some cases these models have been given ontological privilege. What has made contemporary concepts and ideologies of the virtual possible is that these preexisting systems of thought have expanded out of the imagination, and manifested themselves in the development and understanding of technology. The following work, as condensed as it may be, extracts traces of the virtual from past historical and philosophical narratives. These traces show intertextual relationships between seemingly disparate systems of thought that have now been recombined into a working body of "knowledge" under the sign of technology.

I
385 B.C.

This artisan is able to make not only all kinds of furniture but also all plants that grow from the earth, all animals including himself and, besides, the earth and the heavens and the gods, all things in heaven and all things in Hades below the earth.

This program is able to make not only all kinds of furniture but also all plants that grow from the earth, all animals, itself, and, besides, the earth and the heavens and the gods, all things in heaven and all things in Hades below the earth.

II
60 B.C.

There is no visible object that consists of atoms of one kind only. Everything is composed of a mixture of elements. The more qualities and powers a thing possesses, the greater variety it attests in the forms of its component atoms.

There is no visible object that consists of pixels of one kind only. Everything is a recombinant mixture of representation. The more qualities and powers an image possesses, the greater variety it attests in the forms of its component pixels.

III
A.D. 250

Let us then, make a mental picture of our
universe: each member shall remain what it is,
distinctly apart; all is to form, as far as possible,
a complete unity so that whatever comes into
view, say the outer orb of the heavens, shall
bring immediately with it the vision, on the
one plane, of the sun and all of the stars with
earth and sea and all living things as if exhibited upon a transparent globe.

Let us then, make a virtual representation of our
universe: each member shall remain what it is,
distinctly apart; all is to form, as far as possible,
a complete unity so that whatever comes into
view, say the outer orb of the heavens, shall bring
immediately with it the vision, on the one plane,
of the sun and all of the stars with earth and sea
and all living things as if exhibited upon a
transparent globe.

IV
A.D. 413

There are many reprobate mingled with the good, and both are gathered together by the gospel as in a dragnet; and in this world, as in a sea, both swim enclosed without distinction in the net.

There are many reprobate mingled with the good, and both are gathered together in the data base as in a dragnet; and in this world, as in a sea, both swim enclosed without distinction in the electronic net.

V
1259

There are two kinds of contact, that of quantity, and that of power. By the former a body can be touched only by a body; by the latter a body can be touched by an incorporeal reality, which moves that body.

There are two kinds of contact, that of surface, and that of power. By the former a body can be touched only by a body; by the latter a body can be touched by an incorporeal reality, which moves that body.

VI
1321

So here on earth, across a slant of light
that parts the air within the sheltering shade
man's arts and crafts contrive, our mortal sight

observes bright particles of matter ranging
up, down, aslant, darting or eddying;
longer and shorter; but forever changing.

So here on screen, across a slant of light
that parts the air within the sheltering shade
man's arts and crafts contrive, our mortal sight

observes bright particles of matter ranging
up, down, aslant, darting or eddying;
longer and shorter; but forever changing.

Welcome

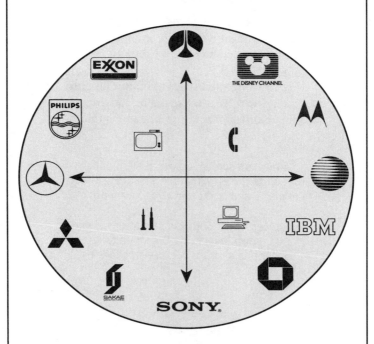

To a World
without Borders

2

Nomadic Power
and Cultural Resistance

The term that best describes the present social condition is liquescence. The once unquestioned markers of stability, such as God or Nature, have dropped into the black hole of scepticism, dissolving positioned identification of subject or object. Meaning simultaneously flows through a process of proliferation and condensation, at once drifting, slipping, speeding into the antinomies of apocalypse and utopia. The location of power—and the site of resistance—rest in an ambiguous zone without borders. How could it be otherwise, when the traces of power flow in transition between nomadic dynamics and sedentary structures—between hyperspeed and hyperinertia? It is perhaps utopian to begin with the claim that resistance begins (and ends?) with a Nietzschean casting-off of the yoke of catatonia inspired by the postmodern condition, and yet the disruptive nature of consciousness leaves little choice.

Treading water in the pool of liquid power need not be an image of acquiescence and complicity. In spite of their awkward situation, the political activist and the cultural activist (anachronistically known as the artist) can still produce disturbances. Although such action may more closely resemble the gestures of a drowning person, and it is uncertain just what is being disturbed, in this situation the postmodern roll of the dice favors the act of disturbance. After all, what other chance is there? It is for this reason that former strategies of "subversion" (a word which in critical discourse has about as much meaning as the word "community"), or camouflaged attack, have come under a cloud of suspicion. Knowing what to subvert assumes that forces of oppression are stable and can be identified and separated— an assumption that is just too fantastic in an age of dialectics in ruins. Knowing how to subvert presupposes an understanding of the opposition that rests in the realm of certitude, or (at least) high probability. The rate at which strategies of subversion are co-opted indicates that the adaptability of power is too often underestimated; however, credit should be given to the resisters, to the extent that the subversive act or product is not co-optively reinvented as quickly as the bourgeois aesthetic of efficiency might dictate.

The peculiar entwinement of the cynical and the utopian in the concept of disturbance as a necessary gamble is a heresy to those who still adhere to 19th-century narratives in which the mechanisms and class(es) of oppression, as well as the tactics needed to overcome them, are clearly identified. After all, the wager is deeply connected to conservative apologies for Christianity, and the attempt to appropriate rationalist rhetoric and models to persuade the fallen to return to traditional eschatology. A renounced Cartesian like Pascal, or a renounced revolutionary like Dostoyevsky, typify its use. Yet it must be realized that the promise of a better future, whether secular or spiritual, has always presupposed the economy of the wager. The connection between history and necessity is cynically humorous when one looks back over the trail of political and cultural debris of revolution and near-revolution in ruins. The French revolutions from 1789 to 1968 never stemmed the obscene tide of the commodity (they seem to have helped pave the way), while the Russian and Cuban revolutions merely replaced the commodity with the totalizing anachronism of the bureaucracy. At best, all that is derived from these disruptions is a structure for a nostalgic review of reconstituted moments of temporary autonomy.

The cultural producer has not fared any better. Mallarmé brought forth the concept of the wager in *A Roll of the Dice*, and perhaps unwittingly liberated invention from the bunker of transcendentalism that he hoped to defend, as well as releasing the artist from the myth of the poetic subject. (It is reasonable to suggest that de Sade had already accomplished these tasks at a much earlier date). Duchamp (the attack on essentialism), Cabaret Voltaire (the methodology of random production), and Berlin dada (the disappearance of art into political action) all disturbed the cultural waters, and yet opened one of the cultural passages for the resurgence of transcendentalism in late Surrealism. By way of reaction to the above three, a channel was also opened for formalist domination (still to this day the demon of the culture-text) that locked the culture-object into the luxury

market of late capital. However, the gamble of these fore-
runners of disturbance reinjected the dream of autonomy
with the amphetamine of hope that gives contemporary
cultural producers and activists the energy to step up to the
electronic gaming table to roll the dice again.

In *The Persian Wars*, Herodotus describes a feared people known as
the Scythians, who maintained a horticultural-nomadic
society unlike the sedentary empires in the "cradle of civilization." The homeland of the Scythians on the Northern
Black Sea was inhospitable both climatically and geo-
graphically, but resisted colonization less for these natural
reasons than because there was no economic or military
means by which to colonize or subjugate it. With no fixed
cities or territories, this "wandering horde" could never
really be located. Consequently, they could never be put on
the defensive and conquered. They maintained their autonomy through movement, making it seem to outsiders
that they were always present and poised for attack even
when absent. The fear inspired by the Scythians was quite
justified, since they were often on the military offensive,
although no one knew where until the time of their instant
appearance, or until traces of their power were discovered.
A floating border was maintained in their homeland, but
power was not a matter of spatial occupation for the Scythians.
They wandered, taking territory and tribute as needed, in
whatever area they found themselves. In so doing, they
constructed an invisible empire that dominated "Asia" for
twenty-seven years, and extended as far south as Egypt. The
empire itself was not sustainable, since their nomadic nature
denied the need or value of holding territories. (Garrisons
were not left in defeated territories). They were free to
wander, since it was quickly realized by their adversaries that

even when victory seemed probable, for practicality's sake it was better not to engage them, and to instead concentrate military and economic effort on other sedentary societies—that is, on societies in which an infrastructure could be located and destroyed. This policy was generally reinforced, because an engagement with the Scythians required the attackers to allow themselves to found by the Scythians. It was extraordinarily rare for the Scythians to be caught in a defensive posture. Should the Scythians not like the terms of engagement, they always had the option of remaining invisible, and thereby preventing the enemy from constructing a theater of operations.

This archaic model of power distribution and predatory strategy has been reinvented by the power elite of late capital for much the same ends. Its reinvention is predicated upon the technological opening of cyberspace, where speed/absence and inertia/presence collide in hyperreality. The archaic model of nomadic power, once a means to an unstable empire, has evolved into a sustainable means of domination. In a state of double signification, the contemporary society of nomads becomes both a diffuse power field without location, and a fixed sight machine appearing as spectacle. The former privilege allows for the appearance of global economy, while the latter acts as a garrison in various territories, maintaining the order of the commodity with an ideology specific to the given area.

Although both the diffuse power field and the sight machine are integrated through technology, and are necessary parts for global empire, it is the former that has fully realized the Scythian myth. The shift from archaic space to an electronic network offers the full complement of nomadic power advantages: The militarized nomads are always on the offensive. The obscenity of spectacle and the terror of speed are their constant companions. In most cases sedentary populations submit to the obscenity of spectacle, and contentedly pay the tribute demanded, in the form of labor, material, and profit. First world, third world, nation or tribe, all must give tribute. The differentiated and hierarchical nations, classes, races, and genders of sedentary modern society all blend under nomadic domination into the role of its service workers—into caretakers of the cyberelite. This separation, mediated by spectacle, offers tactics that are beyond the archaic nomadic model. Rather than a hostile plundering of an adversary, there is a friendly pillage, seductively and ecstatically conducted against the passive. Hostility from the oppressed is rechanneled into the bureaucracy, which misdirects antagonism away from the nomadic power field. The retreat into the invisibility of nonlocation prevents those caught in the panoptic spatial lock-down from defining a site of resistance (a theater of operations), and they are instead caught in a historical tape loop of resisting the monuments of dead capital. (Abortion rights? Demonstrate on the steps of the Supreme Court. For the release of drugs which slow the development of HIV, storm the NIH). No longer needing to take a defensive posture is the nomads' greatest strength.

As the electronic information-cores overflow with files of electronic people (those transformed into credit histories, consumer types, patterns and tendencies, etc.), electronic research, electronic money, and other forms of information power, the nomad is free to wander the electronic net, able to cross national boundaries with minimal resistance from national bureaucracies. The privileged realm of electronic space controls the physical logistics of manufacture, since the release of raw materials and manufactured goods requires electronic consent and direction. Such power must be relinquished to the cyber realm, or the efficiency (and thereby the profitability) of complex manufacture, distribution, and consumption would collapse into a communication gap. Much the same is true of the military; there is cyberelite control of information resources and dispersal. Without command and control, the military becomes immobile, or at best limited to chaotic dispersal in localized space. In this manner all sedentary structures become servants of the nomads.

The nomadic elite itself is frustratingly difficult to grasp. Even in 1956, when C. Wright Mills wrote *The Power Elite*, it was clear that the sedentary elite already understood the importance of invisibility. (This was quite a shift from the looming spatial markers of power used by the feudal aristocracy). Mills found it impossible to get any direct information on the elite, and was left with speculations drawn from questionable empirical categories (for example, the social register). As the contemporary elite moves from centralized urban areas to decentralized and deterritorialized cyberspace, Mills' dilemma becomes increasingly aggravated. How can a subject be critically assessed that cannot be located, examined, or even seen? Class analysis reaches a point of exhaustion. Subjectively there is a feeling of oppression, and yet it is difficult to locate, let alone assume, an oppressor. In all likelihood, this group is not a class at all—that is, an aggregate of people with common political and economic interests—but a downloaded elite military consciousness. The cyberelite is now a transcendent entity that can only be imagined. Whether they have integrated programmed motives is unknown. Perhaps so, or perhaps their predatory actions fragment their solidarity, leaving shared electronic pathways and stores of information as the only basis of unity. The paranoia of imagination is the foundation for a thousand conspiracy theories—all of which are true. Roll the dice.

The development of an absent and potentially unassailable nomadic power, coupled with the rear vision of revolution in ruins, has nearly muted the contestational voice. Traditionally, during times of disillusionment, strategies of retreatism begin to dominate. For the cultural producer, numerous examples of cynical participation populate the landscape of resistance. The experience of Baudelaire comes to mind. In 1848 Paris he fought on the barricades, guided by the notion that "property is theft," only to turn to cynical nihilism after the revolution's failure. (Baudelaire was never able to completely surrender. His use of plagiarism as an inverted colonial strategy forcefully recalls the notion that property is theft). André Breton's early surrealist project—synthesizing the liberation of desire with the liberation of the worker—unraveled when faced with the rise of fascism. (Breton's personal arguments with Louis Aragon over the function of the artist as revolutionary agent should also be noted. Breton never could abandon the idea of poetic self as a privileged narrative). Breton increasingly embraced mysticism in the 30s, and ended by totally retreating into transcendentalism. The tendency of the disillusioned cultural worker to retreat toward introspection to sidestep the Enlightenment question of "What is to be done with the social situation in light of sadistic power?" is the representation of life through denial. It is not that interior liberation is undesirable and unnecessary, only that it cannot become singular or privileged. To turn away from the revolution of everyday life, and place cultural resistance under the authority of the poetic self, has always led to cultural production that is the easiest to commodify and bureaucratize.

From the American postmodern viewpoint, the 19th-century category of the poetic self (as delineated by the Decadents, the Symbolists, the Nabis School, etc.) has come to represent complicity and acquiescence when presented as pure. The culture of appropriation has eliminated this option in and of itself. (It still has some value as a point of intersection. For example, bell hooks uses it well as an entrance point to other discourses). Though in need of revision, Asger Jorn's modernist motto "The avant-garde never gives up!" still has some relevance. Revolution in ruins and the labyrinth of appropriation have emptied the comforting certitude of the dialectic. The Marxist watershed, during which the means of oppression had a clear identity, and the route of resistance was unilinear, has disappeared into the void of scepticism. However, this is no excuse for surrender. The ostracized surrealist, Georges Bataille, presents an option still not fully explored: In everyday life, rather than confronting the aesthetic of utility, attack from the rear through the nonrational economy of the perverse and sacrificial. Such a strategy offers the possibility for intersecting exterior and interior disturbance.

The significance of the movement of disillusionment from Baudelaire to Artaud is that its practitioners imagined sacrificial economy. However, their conception of if was too often limited to an elite theater of tragedy, thus reducing it to a resource for "artistic" exploitation. To complicate matters further, the artistic presentation of the perverse was always so serious that sites of application were often consequently overlooked. Artaud's stunning realization that the body without organs had appeared, although he seemed uncertain as to what it might be, was limited to tragedy and apocalypse. Signs and traces of the body without organs appear throughout mundane experience. The body without organs is Ronald McDonald, not an esoteric aesthetic; after all, there is a critical place for comedy and humor as a means of resistance. Perhaps this is the Situationist International's greatest contribution to the postmodern aesthetic. The dancing Nietzsche lives.

In addition to aestheticized retreatism, a more sociological variety appeals to romantic resisters—a primitive version of nomadic disappearance. This is the disillusioned retreat to fixed areas that elude surveillance. Typically, the retreat is to the most culturally negating rural areas, or to deterritorialized urban neighborhoods. The basic principle is to achieve autonomy by hiding from social authority. As in band societies whose culture cannot be touched because it cannot be found, freedom is enhanced for those participating in the project. However, unlike band societies, which emerged within a given territory, these transplanted communities are always susceptible to infections from spectacle, language, and even nostalgia for former environments, rituals, and habits. These communities are inherently unstable (which is not necessarily negative). Whether these communities can be transformed from campgrounds for the disillusioned and defeated (as in late 60s-early 70s America) to effective bases for resistance remains to be seen. One has to question, however, whether an effective sedentary base of resistance will not be quickly exposed and undermined, so that it will not last long enough to have an effect.

Another 19th-century narrative that persists beyond its natural life is the labor movement—i.e., the belief that the key to resistance is to have an organized body of workers stop production. Like revolution, the idea of the union has been shattered, and perhaps never existed in everyday life. The ubiquity of broken strikes, give-backs, and lay-offs attests that what is called a union is no more than a labor bureaucracy. The fragmentation of the world—into nations, regions, first and third worlds, etc., as a means of discipline by nomadic power—has anachronized national labor movements. Production sites are too mobile and management techniques too flexible for labor action to be effective. If labor in one area resists corporate demands, an alternative labor pool is quickly found. The movement of Dupont's and General Motors' production plants into Mexico, for example, demonstrates this nomadic ability. Mexico as labor colony also allows reduction of unit cost, by eliminating first world "wage standards" and employee benefits. The speed of the corporate world is paid for by the intensification of exploitation; sustained fragmentation of time and of space makes it possible. The size and desperation of the third world labor pool, in conjunction with complicit political systems, provide organized labor no base from which to bargain.

The Situationists attempted to contend with this problem by rejecting the value of both labor and capital. All should quit work—proles, bureaucrats, service workers, everyone. Although it is easy to sympathize with the concept, it presupposes an impractical unity. The notion of a general strike was much too limited; it got bogged down in national struggles, never moving beyond Paris, and in the end it did little damage to the global machine. The hope of a more elite strike manifesting itself in the occupation movement was a strategy that was also dead on arrival, for much the same reason.

The Situationist delight in occupation is interesting to the extent that it was an inversion of the aristocratic right to property, although this very fact makes it suspect from its inception, since even modern strategies should not merely seek to invert feudal institutions. The relationship between occupation and ownership, as presented in conservative social thought, was appropriated by revolutionaries in the first French revolution. The liberation and occupation of the Bastille was significant less for the few prisoners released, than to signal that obtaining property through occupation is a double-edged sword. This inversion made the notion of property into a conservatively viable justification for genocide. In the Irish genocide of the 1840s, English landowners realized that it would be more profitable to use their estates for raising grazing animals than to leave the tenant farmers there who traditionally occupied the land. When the potato blight struck, destroying the tenant farmers' crops and leaving them unable to pay rent, an opening was perceived for mass eviction. English landlords requested and received military assistance from London to remove the farmers and to ensure they did not reoccupy the land. Of course the farmers believed they had the right to be on the land due to their long-standing occupation of it, regardless of their failure to pay rent. Unfortunately, the farmers were transformed into a pure excess population since their right to property by occupation was not recognized. Laws were passed denying them the right to immigrate to England, leaving thousands to die without food or shelter in the Irish winter. Some were able to immigrate to the US, and remained alive, but only as abject refugees. Meanwhile, in the

US itself, the genocide of Native Americans was well underway, justified in part by the belief that since the native tribes did not own land, all territories were open, and once occupied (invested with sedentary value), they could be "defended." Occupation theory has been more bitter than heroic.

In the postmodern period of nomadic power, labor and occupation movements have not been relegated to the historical scrap heap, but neither have they continued to exercise the potency that they once did. Elite power, having rid itself of its national and urban bases to wander in absence on the electronic pathways, can no longer be disrupted by strategies predicated upon the contestation of sedentary forces. The architectural monuments of power are hollow and empty, and function now only as bunkers for the complicit and those who acquiesce. They are secure places revealing mere traces of power. As with all monumental architecture, they silence resistance and resentment by the signs of resolution, continuity, commodification, and nostalgia. These places can be occupied, but to do so will not disrupt the nomadic flow. At best such an occupation is a disturbance that can be made invisible through media manipulation; a particularly valued bunker (such as a bureaucracy) can be easily reoccupied by the postmodern war machine. The electronic valuables inside the bunker, of course, cannot be taken by physical measures.

The web connecting the bunkers—the street—is of such little value to nomadic power that it has been left to the underclass. (One exception is the greatest monument to the war machine ever constructed: The Interstate Highway System. Still valued and well defended, that location shows almost no sign of disturbance.) Giving the street to the most alienated of classes ensures that only profound alienation can occur there. Not just the police, but criminals, addicts, and even the homeless are being used as disrupters of public space. The underclass' actual appearance, in conjunction with media spectacle, has allowed the forces of order to construct the hysterical perception that the streets are unsafe, unwholesome, and useless. The promise of safety and familiarity lures hordes of the unsuspecting into privatized public spaces such as malls. The price of this protectionism is the relinquishment of individual sovereignty. No one but the commodity has rights in the mall. The streets in particular and public spaces in general are in ruins. Nomadic power speaks to its followers through the autoexperience of electronic media. The smaller the public, the greater the order.

The avant-garde never gives up, and yet the limitations of antiquated models and the sites of resistance tend to push resistance into the void of disillusionment. It is important to keep the bunkers under siege; however, the vocabulary of resistance must be expanded to include means of electronic disturbance. Just as authority located in the street was once met by demonstrations and barricades, the authority that locates itself in the electronic field must be met with electronic resistance. Spatial strategies may not be key in this endeavor, but they are necessary for support, at least in the case of broad spectrum disturbance. These older strategies of physical challenge are also better developed, while the electronic strategies are not. It is time to turn attention to the electronic resistance, both in terms of the bunker and the nomadic field. The electronic field is an area where little is known; in such a gamble, one should be ready to face the ambiguous and unpredictable hazards of an untried resistance. Preparations for the double-edged sword should be made.

Nomadic power must be resisted in cyberspace rather than in physical space. The postmodern gambler is an electronic player. A small but coordinated group of hackers could introduce electronic viruses, worms, and bombs into the data banks, programs, and networks of authority, possibly bringing the destructive force of inertia into the nomadic realm. Prolonged inertia equals the collapse of nomadic authority on a global level. Such a strategy does not require a unified class action, nor does it require simultaneous action in numerous geographic areas. The less nihilistic could resurrect the strategy of occupation by holding data as hostage instead of property. By whatever means electronic authority is disturbed, the key is to totally disrupt command and control. Under such conditions, all dead capital in the military/corporate entwinement becomes an economic drain—material, equipment, and labor power all would be left without a means of deployment. Late capital would collapse under its own excessive weight.

Even though this suggestion is but a science-fiction scenario, this narrative does reveal problems which must be addressed. Most obvious is that those who have engaged cyberreality are generally a depoliticized group. Most infiltration into cyberspace has either been playful vandalism (as with Robert Morris' rogue program, or the string of PC viruses like Michaelangelo), politically misguided espionage (Markus Hess' hacking of military computers, which was possibly done for the benefit of the KGB), or personal revenge against a particular source of authority. The hacker* code of ethics discourages any act of disturbance in cyberspace. Even the Legion of Doom (a group of young hackers that put the fear into the Secret Service) claims to have never damaged a system. Their activities were motivated by curiosity about computer systems, and belief in free access to information. Beyond these very focused concerns with decentralized information, political thought or action has never really entered the group's consciousness. Any trouble that they have had with the law (and only a few members break the law) stemmed either from credit fraud or electronic trespass. The problem is much the same as politicizing scientists whose research leads to weapons development. It must be asked, How can this class be asked to destabilize or crash its own world? To complicate matters further, only a few understand the specialized knowledge necessary for such action. Deep cyberreality is the least democratized of all frontiers. As mentioned above, cyberworkers as a professional class do not have to be fully unified, but how can enough members of this class be enlisted to stage a disruption, especially when cyberreality is under state-of-the-art self-surveillance?

These problems have drawn many "artists" to electronic media, and this has made some contemporary electronic art so politically charged. Since it is unlikely that scientific or techno-workers will generate a theory of electronic disturbance, artists-activists (as well as other concerned groups) have been left with the responsibility to help provide a

---

* "Hacker" refers here to a generic class of computer sophisticates who often, but not always, operate counter to the needs of the military/corporate structure. As used here the term includes crackers, phreakers, hackers proper, and cypherpunks.

critical discourse on just what is at stake in the development of this new frontier. By appropriating the legitimized authority of "artistic creation," and using it as a means to establish a public forum for speculation on a model of resistance within emerging techno-culture, the cultural producer can contribute to the perpetual fight against authoritarianism. Further, concrete strategies of image/text communication, developed through the use of technology that has fallen through the cracks in the war machine, will better enable those concerned to invent explosive material to toss into the political-economic bunkers. Postering, pamphleteering, street theater, public art—all were useful in the past. But as mentioned above, where is the "public"; who is on the street? Judging from the number of hours that the average person watches television, it seems that the public is electronically engaged. The electronic world, however, is by no means fully established, and it is time to take advantage of this fluidity through invention, before we are left with only critique as a weapon.

Bunkers have already been described as privatized public spaces which serve various particularized functions, such as political continuity (government offices or national monuments), or areas for consumption frenzy (malls). In line with the feudal tradition of the fortress mentality, the bunker guarantees safety and familiarity in exchange for the relinquishment of individual sovereignty. It can act as a seductive agent offering the credible illusion of consumptive choice and ideological peace for the complicit, or it can act as an aggressive force demanding acquiescence for the resistant. The bunker brings nearly all to its interior with the exception of those left to guard the streets. After all, nomadic power does not offer the choice not to work or not to consume. The bunker is such an all-embracing feature of everyday life that even the most resistant cannot always approach it critically. Alienation, in part, stems from this uncontrollable entrapment in the bunker.

Bunkers vary in appearance as much as they do in function. The nomadic bunker—the product of "the global village"—has both an electronic and an architectural form. The electronic form is witnessed as media; as such it attempts to colonize the private residence. Informative distraction flows in an unceasing stream of fictions produced by Hollywood, Madison Avenue, and CNN. The economy of desire can be safely viewed through the familiar window of screenal space. Secure in the electronic bunker, a life of alienated autoexperience (a loss of the social) can continue in quiet acquiescence and deep privation. The viewer is brought to the world, the world to the viewer, all mediated through the ideology of the screen. This is virtual life in a virtual world.

Like the electronic bunker, the architectural bunker is another site where hyperspeed and hyperinertia intersect. Such bunkers are not restricted to national boundaries; in fact, they span the globe. Although they cannot actually move through physical space, they simulate the appearance of being everywhere at once. The architecture itself may vary considerably, even in terms of particular types; however, the logo or totem of a particular type is universal, as are its consumables. In a general sense, it is its redundant participation in these characteristics that make it so seductive.

This type of bunker was typical of capitalist power's first attempt to go nomadic. During the Counterreformation, when the Catholic Church realized during the Council of Trent (1545-63) that universal presence was a key to power in the age of colonization, this type of bunker came of age. (It took the full development of the capitalist system to produce the technology necessary to return to power through absence). The appearance of the church in frontier areas both East and West, the universalization of ritual, the maintenance of relative grandeur in its architecture, and the ideological marker of the crucifix, all conspired to present a reliable place of familiarity and security. Wherever a person was, the homeland of the church was waiting.

In more contemporary times, the gothic arches have transformed themselves into golden arches. McDonalds' is global. Wherever an economic frontier is opening, so is a McDonalds'. Travel where you might, that same hamburger and coke are waiting. Like Bernini's piazza at St. Peters, the golden arches reach out to embrace their clients—so long as they consume, and leave when they are finished. While in the bunker, national boundaries are a thing of the past, in fact you are at home. Why travel at all? After all, wherever you go, you are already there.

There are also sedentary bunkers. This type is clearly nationalized, and hence is the bunker of choice for governments. It is the oldest type, appearing at the dawn of complex society, and reaching a peak in modern society with conglomerates of bunkers spread throughout the urban sprawl. These bunkers are in some cases the last trace of centralized national power (the White House), or in others, they are locations to manufacture a complicit cultural elite (the university), or sites of manufactured continuity (historical monuments). These are sites most vulnerable to electronic disturbance, as their images and mythologies are the easiest to appropriate.

In any bunker (along with its associated geography, territory, and ecology) the resistant cultural producer can best achieve disturbance. There is enough consumer technology available to at least temporarily reinscribe the bunker with image and language that reveal its sacrificial intent, as well as the obscenity of its bourgeois utilitarian aesthetic. Nomadic power has created panic in the streets, with its mythologies of political subversion, economic deterioration, and biological infection, which in turn produce a fortress ideology, and hence a demand for bunkers. It is now necessary to bring panic into the bunker, thus disturbing the illusion of security and leaving no place to hide. The incitement of panic in all sites is the postmodern gamble.

VII
1500

Of dreaming. It shall seem to men that they see destructions in the sky, and flames descending therefrom shall seem to fly away in terror; they shall hear creatures of every kind speaking human language; they shall run in a moment to diverse parts of the world without movement; they shall see the most radiant splendors amidst darkness.

Of dreaming. It shall seem to men that they experience destructions in the sky and flames descending therefrom shall seem to fly away in terror; they shall hear creatures of every kind speaking human language; they shall travel in a moment to diverse parts of the world without movement; they shall see the most radiant splendors amidst darkness.

VIII
1641

Nothing conduces more to the obtaining of a secure knowledge of reality than a previous accustoming of ourselves to entertain doubts especially about corporeal things.

Nothing conduces more to the obtaining of an uncensored knowledge of reality than a previous accustoming of ourselves to entertain doubts especially about corporeal things.

Hence, at least through the instrumentality of the Divine power, mind can exist apart from body, and body apart from mind.

Hence, at least through the instrumentality of the Virtual power, mind can exist apart from body, and body apart from mind.

This is not a

Ceci n'est pas une pipe.

pipe
either.

3

Video and Resistance:

Against Documentaries

The medium of video was born in crisis. This postmodern technology
has been shoved back into the womb of history with the
demand that it progress through the same developmental
stages as its older siblings, film and photography. The
documentary—the paramount model for resistant video
production—gives witness less to the endless parade of
guerrilla actions, street demonstrations, and ecological di-
sasters than it does to the persistence of Enlightenment
codes of truth, knowledge, and a stable empirical reality.
The hegemony of the documentary moves the question of
video technology away from its function as a simulator, and
back to a retrograde consideration of the technology as a
replicator (witness). Clearly technology will not save us
from the insufferable condition of eternal recurrence.

Recall file entitled "Enlightenment." Enlightenment: A historical
moment past, which must now be looked upon through the
filter of nostalgia. Truth was so simple then. The senses were
trusted, and the discrete units of sensation contained knowledge. To those ready to observe, nature surrendered its
secrets. Every object contained useful pieces of data explod-
ing with information, for the world was a veritable network
of interlocking facts. Facts were the real concern: everything observable was endowed with facticity. Everything
concrete merited observation, from a grain of sand to social
activity. "Knowledge" went nova. The answer to the problem of managing geometrically cascading data was
specialization: Split the task of observation into as many
categories and subcategories as possible to prevent observa-
tional integrity from being distracted by the proliferation of
factual possibility. (It is always amazing to see authoritarian
structures run wild in the utopian moment). Specialization
worked in the economy (complex manufacture) and in
government management (bureaucracy); why not also with
knowledge? Knowledge entered the earthly domain (as
opposed to the transcendental), giving humanity control
over its own destiny and initiating an age of progress with
science as redeemer.

In the midst of this jubilation, a vicious scepticism haunted
the believers like the Encyclopedists, the new social think-
ers (such as Turgot, Fontenelle, and Condorcet), and later,
the logical positivists. The problem of scepticism was exem-
plified by David Hume's critique of the empirical model,
which placed Enlightenment epistemology outside the realm
of certainty. The senses were shown to be unreliable con-
veyers of information, and factual associations were revealed
as practical inference. Strengthened by the romantic cri-

tique developed later under the banner of German Idealism, the argument became acceptable that the phenomenal world was not a source of knowledge, since perception could be structured by given mental categories which might or might not show fidelity to a thing-in-itself. Under this system, science was reduced to a practical mapping of spatial-temporal constellations. Unfortunately, the idealists were unable to escape the scepticism from which they had emerged. Their own system of transcendentalism was just as susceptible to the sceptic's arguments.

Science found itself in a peculiar position in regard to the 19th-century sociology of knowledge. Since it did produce what secularists interpreted as desirable practical results, it became an ideological legitimizer even on the ordinary level of everyday life. Within the sceptic's vacuum, empirical science by default usurped the right to pronounce what was real in experience. Sensible judgment was secure in the present, but to judge past events required immediate perception to be reconstituted through memory. The problem of memory was transformed into a technological problem because the subjective elements of memory led to the decay of the facticity of the sensible object, and written representation as a means to maintain history was insufficient. Although theory and method were mature and legitimized, a satisfactory technology had yet to emerge. This problem finally resolved itself with the invention of photography. Photography could provide a concrete visual record (vision being the most trustworthy of the senses) as an account of the past. Photography represented facts, rather than subjectively dissolving them into memory, or abstracting them as with writing. At last, there was a visual replicator to produce a record independent of the witness. Technology could mediate perception, and thereby impose objectivity upon the visual record. To this extent, photography was embraced more as a scientific tool than as a means to manifest aesthetic intent.

Artists from all media began to embrace the empirical model, which had been rejuvenated by these innovations in replicating technology. Their interest in turn gave birth to Realism and literary Naturalism. In these new genres, the desire for replication became more complex. A new political agenda had insinuated itself into cultural production. Unlike in the past when politics generally served to maintain the status quo, the agenda of the newly-born left began to make a clear-cut appearance in empirical cultural representation. The proponents of this movement no longer worshipped the idealistic cultural icons of the romantic predecessors, but fetishized facticity—tendencies that reduced the artist's role to that of mechanical reproduction. The visual presentation of factual data allowed one to objectively witness the injustice of history, providing those eliminated from the historical record a way to make their places known. The use of traditional media combined with Enlightenment epistemology to promote a new leftist ideology that failed relatively fast. Even the experimental novels of Zola, in the end, could only be perceived as fiction, not as historical accounts. The Realist painters' work seemed equally unreliable, as the paintbrush was not a satisfactory technological means to insure objectivity, while its product was tied too closely to an elitist tradition and to its institutions. Perhaps their only actual victory was to produce a degraded sign of subversive intent that meekly insisted on the horizontalization of traditional aesthetic categories, particularly in the area of subject matter.

By the end of the century, having nowhere else to turn, some leftist cultural producers began to rethink photography and its new advancement, film. The first documentary makers intended to produce an objective and accurate visual record of social injustice and leftist resistance, and guided by those aims the documentary began to take form. The excitement over new possibilities for socially responsible representation allowed production to precede critical reflection about the medium, and the mistakes that were made continue as institutions into the present.

The film documentary was a catastrophe from its inception. Even as far back as the Lumière brothers' work, the facticity of nonfiction film has been crushed under the burden of ideology. A film such as *Workers Leaving the Lumière Factory* functions primarily as an advertisement for industrialization—a sign of the future divorced from the historical forces which generated it. In spite of its static camera and the necessary lack of editing, the function of replication was lost, because the life presented in the film was yet to exist for most. From this point on, the documentary proceeded deeper into its own fatality. A film such as *Elephant Processions at Phnom Penh* became the predecessor of what we now think of as the cynical postmodern work. The documentary went straight to the heart of colonial appropriation. This film was a spectacular sideshow that allowed the viewer to temporarily enter a culture that never existed. It was an opportunity to revel in a simulated event, again isolated from any type of historical context. In this sense, Lumière was Disney's predecessor. Disney World is the completion of the Lumière cultural sideshow project. By appropriating cultural debris and reassembling it in a means palatable for temporary consumption, Disney does in 3-D what Lumière had done in 2-D: produce a simulation of the world culture-text in the fixed location of the bunker.

The situation continued to worsen. Robert Flaherty introduced complex narrative into the documentary in his film *Nanook of the North*. The film was marked by an overcoded film grammar that transcendentally generated a story out of what were supposed to be raw facts. The gaps between the disparate re-presented images had to be brought together by the glue of the romantic ideology favored by the filmmaker. In a manner of speaking, this had to happen, since there were no facts to begin with, but only reconstituted memory. Flaherty's desire to produce the exotic led him to simulate a past that never existed. In the film's most famous sequence, Flaherty recreates a walrus hunt. Nanook had never been on a hunt without guns, but Flaherty insisted he use harpoons. Nanook had a memory of what his father had told him about traditional hunting, and he had seen old Eskimo renderings of it. Out of these memories, entwined with Flaherty's romantic conceptions, the walrus hunt was reenacted. Representation was piled on representation under the pretense of an unachievable originality. It did make an exciting and entertaining story, but it had no more factual integrity than D. W. Griffiths' *Birth of a Nation*.

It is unnecessary to repeat the cynical history of the documentary oscillating along the political continuum from Vertov to Riefenstahl. In all cases it has been fundamentally cynical—a political commodity doomed by the very nature of the technology to continually replay itself within the economy of desire. Film is not now nor has it ever been the technology of truth. It lies at a speed of 24 frames a second. Its value is not as a recorder of history, but simply as a means of communication, a means by which meaning is generated. The frightening aspect of the documentary film is that it can generate rigid history in the present in the same manner that Disney can generate the colonial meaning of the culture of the Other. Whenever imploded films exist simultaneously as fiction and nonfiction they stand as evidence that history is made in Hollywood.

The documentary's uneasy alliance with scientific methodology attempts to exploit the seeming power of science to stop the drift of multifaceted interpretation. Justifiably or not, scientific evidence is incontrovertible; it rests comfortably under the sign of certitude. This is the authority that the documentary attempts to claim for itself. Consequently, documentary makers have always used authoritarian coding systems to structure the documentary narrative.

This strategy relies primarily on the complete exhaustion of the image at the moment of immediate apprehension. The narrative structure must envelop the viewer like a net and close off all other possible interpretations. The narrative guiding the interpretation of the images must flow along a unilinear pathway, at such a speed that the viewer has no time for any reflection. Key in this movement is to produce the impression that each image is causatively linked to the images preceding it. Establishment of causality between the images renders a seamless effect and keeps the viewers' interpretive flow moving along a predetermined course. The course ends with the conclusion prepared by the documentary maker in constructing the causal chain of images, offering what seems to be an incontrovertible resolving statement. After all, who can challenge replicated causality? Its legitimation by traditional rational authority is too great. A documentary fails when the causal chain breaks down, showing the seams and allowing a moment of disbelief to disrupt the predetermined interpretive matrix. Without the scientific principle of causality rigorously structuring the narrative, the documentary's legitimized authority dissipates quite rapidly, revealing its true nature as fictional propaganda. When a legitimation crisis occurs in the film, the image becomes transparent, rather than exhausting itself, and the ideology of the narrative is displayed in all its horrifying glory. The quality documentary does not reveal itself, and it is this illusionistic chicanery—first perfected by Hollywood realism—that unfortunately guides the grand majority of documentary and video witness work that leftist cultural workers currently produce in endless streams.

This pitiful display is particularly insidious because it turns the leftist cultural workers into that which they most fear: Validators of the conservative interpretive matrix. If the fundamental principle of conservative politics is to maintain order for the sake of economy, to complement the needs and desires of the economic elite, and to discourage social heterogeneity, then the documentary, as it now stands, is complicit in participating in that order, even if it flies the banner of social justice over its ideological fortress. This is true because the documentary does not create an opportunity for free thought, but instills self-censorship in the viewer, who must absorb its images within the structure of a totalizing narrative. If one examines the sign of censorship itself, as it was embodied, for example, in Jesse Helms' criticisms of Andre Serrano's *Piss Christ*, one can see the methods of totalizing interpretation at work. Helms argued that a figure of Christ submerged in piss leads to a single conclusion, that the work is an obscene sacrilege. Helms' interpretation is a fair one; however, it is not the only one. Helms used senatorial spectacle as an authority to legitimize and totalize his interpretation. Under his privileged interpretive matrix, the image is immediately exhausted. However, anyone who reflects on Serrano's image for only a moment can see that numerous other meanings are contained within it. There are meanings that are both critical and aesthetic (formal). Helms' overall strategy was not so much to use personal power as a means to censorship, but to create the preconditions for the public to blindly follow into self-censorship, thereby agreeing to the homogenous order desired by the elite class. The resistant documentary depends upon this same set of conditions for its success. The long-term consequences of using such methods, even with good intentions, is to make the viewer increasingly susceptible to illusionistic narrative structure, while the model itself becomes increasingly sophisticated through its constant revision. Anywhere along the political continuum the electronic consumer turns, s/he is treated like media sheep. To stop this manipulation, documentary makers must refuse to sacrifice the subjectivity of the viewer. The nonfiction film needs to travel other avenues than the one inherited from tradition.

Planning a generic leftist documentary for PBS. Subject: The guerrilla war in ___?___ (choose a third-world nation).

1. Choose a title carefully, since it is one of the primary framing devices. It should present itself purely as a description of the images contained in the work, but should also function as a privileged ideological marker. For example, "The Struggle for Freedom in _____." Remember, do not mention "guerrillas" in the title. Such words have a conno-

tation of a lost or subversive cause that could lead to irrational violent action, and that scares liberals.

2. If you have a large enough budget (and you probably do if you are making yet another film on political strife), open with a lyrical aerial shot of the natural surroundings of the country in question. Usually the countryside is held by the guerrillas. This is good. You now have the traditional authority of nature (and the morality of the town/country distinction) on your side. These are two foundational codes of didactic western art. They are rarely questioned, and will create a channel leading the viewer to the belief that you are filming a populist uprising.

3. Dissolve to the particular band of guerrillas that you are going to film. Do not show large armies, and show only small arms, not heavy weaponry. Remember, the guerrillas must look like real underdogs. Americans love that code. If you must talk about the size of the rebel army (for instance, to show the amount of popular support for the resistance), keep it abstract; give only the statistics. Large military formations have that Nuremberg look to them. If at all possible, choose a band comprised of families: It shows real desperation when an entire extended family is fighting. Keep in mind that one of your key missions is to humanize the rebels while making the dominant group an evil abstraction. Finish this sequence by stylishly introducing each of the rebels as individuals.

4. For the next sequence, single out a family to represent the group. Interview each member. Address their motivations for resistance. Follow them throughout the day. Capture the hardships of rebel activity. Be sure to show the sleeping arrangements and the poverty of the food, but concentrate on

what the fight is doing to the family. End the sequence by showing the family involved in a recreational activity. This will demonstrate the rebels' ability to endure, and to be human in the face of catastrophe. It is also the perfect segue into the next sequence: "In this moment of play, who could have imagined the tragedy that would befall them . . ."

5. Having established the rebels as real, feeling people, it is time to turn to the enemy, by showing for instance an atrocity attributed to them. (Never show the enemy themselves; they must remain an alien abstraction, an unknown to be feared.) It is a preferable if a distant relative of the focus family is killed or wounded in the represented enemy action. Document the mourning of the fellow rebels.

6. With the identities of both the rebels and the enemy established, you must now show an actual guerrilla action. It should be read as a defensive maneuver with no connotation of vengeance. Make sure that it is an evening or morning raid, to lessen sympathy for the enemy as individuals. The low light will keep them hidden and allow the sparks of the return gunfire to represent the enemy as depersonalized. Do not show guerillas taking prisoners: It is difficult to maintain viewers' sympathy for the rebels if they are seen sticking automatic weapons in the backs of the enemy and marching them along. Finally, only show the action if the rebels seem to win the engagement.

7. In the victory sequence it is important to show the tie between the rebels and the nonmilitary personnel of the countryside. With the enemy recently beaten, it is safe to go to town and celebrate with the agrarian class. You can include speeches and commemorations in this sequence.

Show the peasants giving the rebels food, while the rebels give the civilians nonmilitary materials captured during the raid. But most importantly, ensure that the sequence has a festive spirit. This will add an emotional contrast to the closing sequence.

8. Final sequence: Focus on the rebel group expressing their dreams of victory and vowing never to surrender. This should cap it: You are now guaranteed a sympathetic response from the audience. The sympathy will override any critical reflection, making the audience content to ride the wave of *your* radical subjectivity. Roll credits. Perhaps add a postscript by the filmmaker on how touched and amazed s/he was by the experience.

In creating a documentary, one small adjustment could be made with minimal disturbance to the traditional model—to announce for a given work that the collection of images presented have already been fully digested within a specialized cultural perspective. Make sure the viewers know that they are watching a *version* of the subject matter, not the thing in itself. This will not cure the many ills of documentary film/video, since versions themselves are prepackaged, having little meaning in relation to other versions; however, it would make the documentary model a little less repugnant, since this disclaimer would avoid the assertion that one was showing the truth of the matter. This would allow the system to remain closed, but still produce the realization that what is being documented is not a concrete history, but an independent semiotic frame through which sensation has been filtered and interpreted.

Take, for instance, documentaries on a subject regarded almost universally as pleasant and innocuous, such as nature. It becomes readily apparent that nature itself is not the subject, nor could it be. Rather, the simulation of nature is actually a repository for specialized cultural perspectives and myths that are antithetical to the sign of civilization. Consider the following versions:

1. Aestheticized Nature. This is a viewpoint common to most National Geographic documentaries. In this formulation, nature is presented as the original source of beauty, grandeur, and grace. Even the most violent events become precious aesthetic processes that must be preserved. This is even true in the presentation of "exotic" racial/ethnic groups! The world is reduced to an art museum that testifies to the cosmological and teleological perfection of nature. Nature's highest function is to exist for aesthetic appreciation. Both the aesthetics and the ideology that conjure this beatific version of nature come from a well-packaged nostalgic romanticism that determine both the documentary maker's expectations and the method for filming and editing.

2. Darwinian Nature. This conception of nature is best represented by the series *The Trials of Life*. In this treatment the Hobbesian universe comes alive, and the war of all against all is graphically depicted. This blood-and-guts version of nature assembles the signage of survivalist ideology to re-present the blind gropings of a cold and uncaring universe. It is a remembrance of the fatality of the world prior to the order of civilization. Such work acts as an ideological bunker defending the luxury of order produced by the police state.

3. Anthropomorphic Nature. This interpretation revolves around the question of "How are animals like people?"

Typical of Disney documentaries or television shows such as *Wild Kingdom*, these films are insufferably cute, and present the natural order as one of innocence. This is not surprising, since these presentations are targeted at children, and so the conflation of human beings (particularly children) with animals is regarded as a good rubric for "healthy" socialization. These films concentrate on animals' nurturing behavior and on their modest "adventures," interpreting nature as a bourgeois entity.

In all such readings, the viewer is presented with an artificially constructed pastiche of images that offers only limited possibilities for the mythic establishment of nature. Nature exists as merely a semiotic construction used to justify some ideological structure. Nature as code is kept fresh by showing animals and panoramic landscapes that are then overlaid with ideological interpretive frameworks. Nature films have never documented anything other than the artificial— that is, institutionally-constructed value systems. Much the same can be said about the political documentary, since only the contingent aspects are different. The filmmaker then shows us people and cities, rather than animals and landscapes.

The various versions of the present that the documentary imposes on its viewers are refashioned by the film/video form into electronic monuments sharing a number of characteristics with their architectural counterparts. Typically, leftist documentaries parallel the function of monuments and participate in the spectacle of obscenity to the following extent:

1. Monuments function as concrete signs of an imposed reconstituted memory.

2. Monumentalism is the concrete attempt to halt the proliferation of meaning in regard to the interpretation of convulsive events. Monuments are not the signs of freedom that they appear to be, but the very opposite, signs of imprisonment, quelling freedom of speech, freedom of thought, and freedom of remembrance. As overseers in the panoptic prison of ideology, their demand for submission is masochistically obeyed by too many.

3. The return of cultural continuity is what exalts the monument in the eyes of the complicit. In its cloak of silence, the monument can easily repress contradiction. To those whose values they represent, monuments offer a peaceful space through the familiarity of cynical tradition. At the monument, the complicit are not burdened with alienation arising from diversity of opinion, nor with the anxiety of moral contradiction. They are safe from the disturbance of reflection. Monuments are the ultimate ideological bunkers—the concrete manifestations of fortress mentality.

To be sure, there are differences between the architectural monuments of dominant culture, and the monuments to resistant culture, such as documentaries; those of resistant culture do not aspire to maintain the status quo, nor do they project a false continuity onto the wound of history. The problem is that many of these monuments do aspire to an eventual dominance; they aspire to produce an icon that is above critical examination. Thus far no sacred icons have been intentionally produced through the production of documentaries, but some have been accidentally produced through media spectacle. The most notable examples are the Hill/Thomas hearings, and the Rodney King beating.

Certain images derived from these tapes have transcended the mundane to become sacred images for a broad spectrum of society. Like any sacred image, these icons exhaust themselves on impact, and anyone who insinuates that meanings other than the one that immediately presents itself are layered into the image will be visited with a rain of punishment. These images are so emotionally charged that they produce a panic, motivating a blind and vicious attack on any interpretive heresy. They are to the left very much what the image of the aborted fetus is to the radical right. If autonomy is the goal of resistant image production, the monumentality of the sacred must be eliminated from it.

One practical advantage of reality video (video that appears to replicate history) must be recognized—its function as a democratic form of counter-surveillance. No matter how simple the video technology, it easily becomes seen as a threat. It is perceived as a receptacle for guilt that can instantly replay acts of transgression. As the perfect judicial witness, its objectivity cannot be legally questioned. Yet as an instrument of intimidation against the transgressions of power, video functions only within limited parameters. Its strict rational-legal power operates only in the context of exhausted meaning. It is a useful defense in the legal system and in media spectacle, but it is detrimental to the understanding of media itself, as it promotes the authoritarian aesthetics of exhaustion.

The supremacy of reality video as the model for resistant cultural production must be challenged by those who want to see the medium of video go beyond its traditional function as propaganda, while still maintaining resistant political qualities. To eradicate reality video is unnecessary, but to curb its authority is essential. This goal can be best accomplished by developing a postmodern conceptual structure that blends with video's postmodern techno-structure. The fundamental contradiction of using 18th-century epistemology with 19th-century production techniques is that this will never adequately address the contemporary problems of representation in the society of simulation, just as medieval theology was incapable of addressing the challenges of 17th- and 18th-century philosophy.

To resolve this contradiction, one must abandon the assumption that the image contains and shows fidelity to its referent. This in turn means that one can no longer use the code of causality as a means of image continuity. Preferably, one should use liquid associational structures that invite various interpretations. To be sure, all imaging systems are mediated by the viewer: The question is, to what degree? Few systems invite interpretation, and hence meaning is imposed more often than it is created. Many producers, for fear of allowing interpretation to drift out of control, have shunned the use of associational structures for politicized electronic imaging. Further, associational films tend toward the abstract, and therefore become confusing, making them ineffective among the disinterested. These problems prompt the eternal return to more authoritarian models. The answer to such commentary is that the viewer deserves the right to disinterest, and the freedom to drift. Confusion should be seen as an acceptable aesthetic. The moment of confusion is the precondition for the scepticism necessary for radical thought to emerge. The goals then of resistant nonfiction video are twofold: Either to call attention to and document the sign construction of simulation, or to establish confusion and scepticism so that simulations cannot function.

The associational video is by its very nature recombinant. It assembles and reassembles fragmented cultural images, letting the meanings they generate wander unbounded through the grid of cultural possibility. It is this nomadic quality that distinguishes them from the rigidly bounded recombinant films of Hollywood; however, like them, they rest comfortably in neither the category of fiction nor nonfiction. For the purposes of resistance, the recombinant video offers no resolution; rather, it acts as a data base for the viewer to make h/is own inferences. This aspect of the recombinant film presupposes a desire on the part of the viewer to take control of the interpretive matrix, and construct h/is own meanings. Such work is interactive to the extent that the viewer cannot be a passive participant. S/he must not be spoonfed a particular point of view for a pedagogical purpose. This characteristic often works against popular interaction, since strategies to break the habitual passive consumption of spectacle have not received much attention. What is more unfortunate is that such work is often perceived to be elitist, because its use of the aesthetics of confusion does not *at present* draw popular support. It should be noted that such commentary generally comes from a well-positioned intelligentsia certain of the correctness of its ideology. Its mission is to not to free its converts, but to keep them locked in and defending the bunker of solidified ideology. It is disturbance through liquidation of these structures that resistant nomadic media attempts to accomplish. This cannot be done by producing more electronic monuments, but rather, by an imaginative intervention and critical reflection liberated in an unresolved and uncertain electronic moment.

IX
1667

And with asphaltic slime; broad as the gate,
Deep to the roots of Hell the gathered beach
They fastened, and the mole immense wrought on
Over the foaming Deep high-arched, a bridge,
Of length prodigious, joining to the wall
Immovable of this now fenceless world.

And with asphaltic slime; broad as the gate,
Deep to the roots of Hell the gathered beach
For the silicon chip immense wrought on
Over the foaming Deep high-arched, a bridge,
Of length prodigious, joining to the wall
Immovable of this now fenceless world.

X
1759

The land here was cultivated for pleasure as
well as from necessity; everywhere the useful
had been made pleasant. The roads were
covered, or rather adorned, with beautifully
formed carriages made of lustrous material,
carrying men and women of extraordinary
beauty and swiftly drawn by large red sheep
whose speed surpasses the finest horses of
Andalusia.

The simuscape here was cultivated for pleasure
as well as from necessity; everywhere the useful
had been made pleasant. The conduits were
covered, or rather adorned, with beautifully
formed carriages made of lustrous light, carrying
men and women of extraordinary resolution and
swiftly drawn by large red electrical surges whose
speed surpasses the finest missiles of Andalusia.

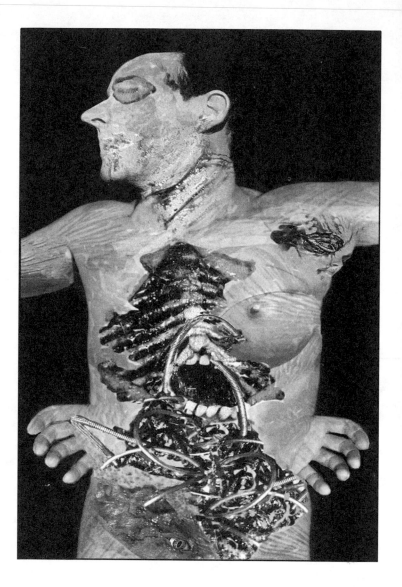

4

The Recombinant Theater
and the Performative Matrix

In some cultures familiar with only modest imaging technologies, people believe that one should not allow oneself to be photographed, as this process steals a part of the soul. This uncanny intuition perhaps shows an understanding that as representation of the self expands, the performative matrix becomes cluttered with simulated persona that can usurp the role of organic self-presentation. The body as representation relinquishes its sovereignty, leaving the image of the body available for appropriation and for reestablishment in sign networks separate from those of the given world. From a contemporary point of view, this is not necessarily negative, since it suggests the possibility that one can continually

Portions of this chapter were originally published in *The Last Sex*. M. and A. Kroker, eds. New York: St. Martins Press, 1993.

reinvent one's character identification and role to better suit one's desires. In light of this possibility, we ought to surrender essentialist notions of self, personality, and body, and take up roles within the dramaturgical grid of everyday life. Yet there is always an uneasiness that accompanies this utopian possibility. This anxiety arises less from the curious nonposition of having no fixed qualities, than it does from the fear that the power of reinvention lies elsewhere. One senses that hostile external forces, rather than self-motivated ones, are constructing us as individuals. This problem becomes increasingly complex in techno-culture, where people find themselves in virtual theaters alien to everyday life but which have a tremendous impact on it. Abstracted representations of self and the body, separate from the individual, are simultaneously present in numerous locations, interacting and recombining with others, beyond the control of the individual and often to h/is detriment. For the critical performer, exploring and interrogating the wanderings and manipulations of the numerous electronic dopplegängers within the many theaters of the virtual should be of primary significance.

Consider the following scenario: A person (P) walks into a bank with the idea of securing a loan. According to the dramaturgical structure of this situation, the person is required to present h/erself as a responsible and trustworthy loan applicant. Being a good performer, and comfortable with this situation, P has costumed h/erself well by wearing clothing and jewelry that indicate economic comfort. P follows the application procedures well, and uses good blocking techniques with appropriate handshakes, standing and sitting as socially expected, and so on. In addition, P has prepared and memorized a well-written script that fully explains h/er need for the loan, as well as h/er ability to repay it. As careful as P is to conform to the codes of the situation, it quickly becomes apparent that h/er performance in itself is not sufficient to secure the loan. All that P has accomplished by the performance is to successfully convince the loan officer to interview h/er electronic double. The loan officer calls up h/er credit history on the computer. It is this body, a body of data, that now controls the stage. It is, in fact, the *only* body which interests the loan officer. P's electronic double reveals that s/he has been late on credit payments in the past, and that she has been in a credit dispute with another bank. The loan is denied; end of performance.

This scenario could just as easily have had a happy ending, but its real importance is to show that the organic performance was primarily redundant. The reality of the applicant was suspect; h/er abstracted image as credit data determined the result of the performance. The engine of the stage, represented by the architecture of the bank, was consumed by the virtual theater. The stage of screenal space, supported by the backstage of data bases and internets, maintains ontological privilege over the theater of everyday life.

With an understanding of the virtual theater, one can easily see just how anachronistic most contemporary performance art is. The endless waves of autoperformance, manifesting themselves as monologues and character bits, serve primarily as nostalgic remembrances of the past, when the performative matrix was centered in everyday life, and focused on organic players. As a work of cultural resistance, the autoperformance's subversive intent appears in its futile attempt to reestablish the subject on the architectural stage. Like most restorationist theater, its cause is dead on arrival.

The performance grid in this situation is already overcoded by the extreme duration of its history, and also suffers from the clutter of codes and simulated persona imposed by spectacle. The attempt to sidestep these problems, by bringing the personal into the discourse, does not have an intersubjective depth of meaning that can maintain itself without networking with coding systems independent of the individual performer. Consequently, the spectacular body and the virtual body consume the personal by imposing their own predetermined interpretive matrices. As shocking as it may sound, the personal is *not* political in recombinant culture.

Case 43

From the notebooks of Jacques Lacan

From the darkness a pre-recorded voice begins to overlap itself in "commentary" on a certain "Case 43" and discussion of the "imaginary status of economic consumption." Then Fon van Voerkom's drawing, "a painful solution," appears on large screen. A few moments later an eye appears on two TV monitors, from which a distorted voice begins to answer the "commentary." The "subject" enters and stands in front of the screen, then begins to make a series of "statements."

The Subject: Born to consume just for the fun of it. Just for the fun of it, mass consumption necessitates self consumption, just for the fun of it. Just for the fun of it auto-cannibalism is the material signifier of excess consumption, just for the fun of it. Just for the fun of it excess consumption is the logic of economic narcissism, just for the fun of it. Just for the fun of it mass consumption equals self-consumption, just for the

Such problems indicate powerfully that the model of production is thoroughly antiquated for performance (as for so much contemporary art). Although in ancient times, the stage was the preeminent platform for the interaction of mythic codes, and although this status remained unquestioned until the 19th century, it has now reached a point of exhaustion. The traditional stage in and of itself is a hollow bunker divorced from power. As a location for disturbance, it offers little hope. Rigor mortis has set in, and what used to be a site for liquid characters, who appeared simply by grabbing a mask, has now become a place where only the

---

fun of it. Auto-cannibalism is the logic of fashion. Deconstruction just for the fun of it. Auto-cannibalism is the praxis of everyday life: I chew my nails just for the fun of it; I eat my hair just for the fun of it; I eat myself just for the fun of it. Consumption is concerned with the internalization of objects, just for the fun of it. Just for the fun of it we consume the objects in order to make them "real," just for the fun of it. Just for the fun of it I eat myself in order to be "real," just for the fun of it. Auto-cannibalism is created just for the fun of it; planned, just for the fun of it; organized through social production, just for the fun of it. We are dogs in love with our own vomit. This is not an aesthetic transgression, this is not a ritual sacrifice, this is not body art, it is only self-consumption, just for the fun of it . . . just for the taste it.

The "Subject" then takes out a razor blade and cuts the palm of his hand. As the blood begins to flow out, the "Subject" drinks the blood for a few moments and then walks away. The "commentary" ends, the large screen image ends, and then the two TV monitors are turned off.

situations of the past or the simulations of the present may be replayed.

Attempts to expand the stage have met with interesting results. The aim of The Living Theater to break the boundaries of its traditional architecture was successful. It collapsed the art and life distinction, which has been of tremendous help by establishing one of the first recombinant stages. After all, only by examining everyday life through the frame of a dramaturgical model can one witness the poverty of this performative matrix. The problem is that effective resistance will not come from the theater of everyday life alone. Like the stage, the subelectronic—in this case the street, in its traditional architectural and sociological form—will have no effect on the privileged virtual stage.

Consider the following scenario: A hacker is placed on stage with a computer and a modem. Working under no fixed time limit, the hacker breaks into data bases, calls up h/er files, and proceeds to erase or manipulate them in accordance with h/er own desires. The performance ends when the computer is shut down.

This performance, albeit oversimplified, signifies the heart of the electronic disturbance. Such an action spirals through the performative network, nomadically interlocking the theater of everyday life, traditional theater, and virtual theater. Multiple representations of the performer all explicitly participate in this scenario to create a new hierarchy of representation. Within the virtual theater, the data structures that contain the electronic representation of the performer are disturbed through their manipulation or deletion. In order for electronic data to act as the reality of a person, the data "facts" cannot be open to democratic manipulation. Data loses privilege once it is found to be invalid or unreliable. This situation offers the resistant performer two strategies: One is to contaminate and call attention to corrupted data, while the other is to pass counterfeit data. Either way, the establishment of the utopian goal of personal reinvention through performative recombination begins to take a form beyond everyday life. Greater freedom in the theater of everyday life can be obtained, once the virtual theater is infiltrated. The liberation gained through the recombinant body can only exist as long as authoritarian codes do not disrupt the performance. For this to happen, the individual must have control of h/er image in all theaters, for only in this way can everyday life performance be aligned with personal desire.

To make the above example more concrete, assume that the hacker is also a female to male cross-dresser. In the performance she accesses h/er identification files, and changes the gender data to "male." S/he leaves the stage, and begins a performance of gender selection on the street. This begins a performance with desire unchained in the theater of everyday life. The gender with which s/he identifies becomes the gender s/he actually is, for no contradictory data resource exists. This performance is not limited to a matter of costuming, but can also affect the flesh. Even biology will begin to collapse. To give an extreme example: Dressed as a man from the waist down, and using "masculine" gesture codes, the performer walks down the street shirtless. S/he is stopped by the police. The appearance of h/er breasts contradicts the desired gender role performance. The police access the electronic information that validates the performer's claim to be a man. The performer is released, since it is not illegal for a man to go shirtless. This performance could easily have gone the other way with the arrest of the performer, but that is extremely unlikely, because such action would require perception to override the data facts.

To say the least, a performance like this is extremely risky. To challenge the codes and unleash desire is generally illegal, particularly as described here. Hacking draws the eye of discipline quickly; it is the best way to destabilize the reality and practical structure of all theaters. Yet these

---

Tongue Spasms

The mouth fragments the body. What remains? A narrow constipation, a violent meaning that makes vomit reason. The grotesque colonization of the oral cavity chews on the silenced body and spits out a bestiality of signs. What remains? Spasms.

The screenal tongue floats freely from its pillars. A sliding surrealistic appendage.

The eye spasms before the virtual tongue, blinding the dominant need for appropriation. What remains after the system digests everything? A nomadic tongue riding the waves of its digital secretions. A post-biological cannibalism that reborders the body. What remains?

The tongue no longer occupies one place.

extreme examples outline the necessary steps needed for a postmodern theater of resistance. Effective performance as a site of resistance must utilize interlocking recombinant stages that oscillate between virtual life and everyday life. This means that the performer must cope with h/er electronic images, and with their techno-matrix. It is time to develop strategies that strike at virtual authority. As yet, there are none. Performers have been too mired in the traditional theater and the theater of everyday life to even realize how the virtual world acts as the theater of final judgment.

The nipple is the matrix of a lost cause, a nostalgia of a network plurality in which one is too few and two is only one possibility. What remains? As screenal tongues cleave and suck the pacifier of unreal ideologies and unreal referents, the cancer of the techno-democracy reveals itself. The nipples mandate the electronic passion of diachronic doubles that blur desire and labor.

Cyber saliva slides in little jerks, punctuating farts and knuckle cracks.

The spasm of digital bytes legitimizes the violence of information. Both the left and right hand are driven by the ritual of representation and sacrifice before the keyboard of dromographic speed. What remains? Hyper-real hands, sociologically unconscious desiring machines, always already possessed. What remains?

The sex speaks of a language based on lubricants, a different kind of saliva.

New theater should tell the viewer how to resist authority, regardless of its source along the political continuum. If we seek liberation through the control of our own images, performance should illustrate resistant processes and explicitly show how to achieve autonomy, however temporary it might be. Self-presentation revealed in the performance must not be perceived by the audience as a self image that should necessarily be copied, as this will end merely as a shift in coding regimes. Rather, one should seek an aesthetics of confusion that reveals potential choices, thus collapsing the bourgeois aesthetic of efficiency.

The virtual tongue fuses with the hot and cold units of pleasure. Unlike things join, tugging sensory hair, and a cannibalism is turned inward. Diseased rumors float back and forth between nano peckers and macro cunts. What remains? A discharge of blind desire moving in and out of virtually gossiping genitals.

Would the virtual tongue multiply and separate toes or simply lick between them?

The big toe is the horror of a base materialism that spasms beyond suitable discourse. Toes lead an ignoble life, seducing the data base with corns, blocking electronic interface with calluses and resisting the drift of information with dirty bunions. What remains after the system digests everything? The ecstatic deformity of pure labor, laughing before the solar anus, flicking mud at the virtual body above it. What remains? The brutal seduction of abandonment more acute in movement.

Already here and yet always one step ahead: It seems that virtual
reality is always about to arrive with the next technological
breakthrough. On the other hand, that curious feeling—
that we are *currently* in a real environment—leads to the
conclusion that virtual reality is located in the near future,
in science fiction, or in an as-yet undeveloped technology.
Perhaps the fact that we are already enveloped by the virtual
is what makes it so unrecognizable. Perhaps it is because a
promise has been issued by technologues, that the boundary
between everyday life and virtual life will soon congeal,
forming completely separate theaters. These promises are

---

*The spasm of the digital body breaks open the orifice of profound
physical impulses.*

The anal night calls the virtual tongue to leave the mouth
and enter it, red and obscene. An eruptive force of luminous
thirst that demands indecent rupture and debauched hacking. What remains? An ontology of farts, of breathless
lacerations that reborder the body and begin to speak. A
revolutionary breakthrough of a post-biological sound. What
remains after the system digests everything? Virtual gas.

what keep the virtual forever invisible. The virtual theater promised by the technologues, like everyday life, will have an enveloping effect. It will be the first engine of the virtual where people will be able to physically interact and have a degree of control over their identities, narrative trajectories, and the objects of interaction. Unlike painting, theater, film, or television, the new virtual theater will make screenal mediation transparent and offer the appearance of unframed experience. This is the idea of virtual reality proper, in its technical sense. However, this technology does not really exist, except in the crudest of forms, and functions primarily as a game. For this reason, the virtual *stage* seems to be nothing worth noting, but as suggested herein, it is already interlocked with everyday life, and already controls the performances of this theater. Should virtual reality proper make its appearance in culture, it must not be confused with virtual power. At present, virtual reality and its promise act as deflectors to turn vision away from the electronic source of domination and authority. The promise of a cybernetic performative matrix serves to alienate us further from our electronic counterparts, falsely leading us to continue believing that electronic bodies do not really exist, let alone that they are signs of authoritarian power. A theater of resistance can be established only if we understand that the virtual world is in the here and now.

The Situationists were correct in their claim that power resides in the spectacle; however, this claim was truer in the past—when the opening shots were fired in the revolution of the economy of desire over the economy of production. Information technology quickly divorced power from the spectacle, and power now wanders invisibly in a cybernetic realm outside of everyday life. Spectacle has become the site of mediation, not so much between social relationships proper, but between the concrete and the virtual worlds, the sedentary and the nomadic, the organic and the electronic, and the present and the absent. To this extent, performance cannot concentrate solely on the virtual. The electronic elements of spectacle are also of great importance and require further investigation, especially since this is the side of the spectacle that mutates at a velocity that parallels consumption. (Architecture and other subelectronic visual markers of the spectacle are not as significant. These forms change too slowly and access to them is limited by geography.) In the electronic image one can detect the clearest traces of the cyberelite, but more importantly, this image is also the source which redistributes identities and lifestyles suitable for excessive consumption. This new social relationship between the electronic body (the body without organs) and the organic body is one of the best resources for performance material. Performance resources must go beyond the organic body, which at present acts as the master link in performative models of representation. In the age of electronic media, it is inappropriate to argue that performance exhausts itself under the sign of the organic. After all, the electronic body is always performing, even if *in absentia* on every stage.

There is every reason to desire the electronic body, and every reason to despise it. This pathological struggle occurs when one views the electronic body, and feelings of sympathy (Husserl) and envy (Benjamin) implode in a schizophrenic moment. As Baudrillard states: "In spite of himself the schizophrenic is open to everything and lives in the most extreme confusion. The schizophrenic is not, as generally claimed, characterized by his loss of touch with reality, but by the absolute proximity to and total instantaneousness with

things, this overexposure to the transparency of the world."
In the debris of intersubjectivity, the organic and the
electronic face each other. The electronic body looks so
real. It moves around, it gazes back, it communicates. Its
appearance is our appearance. Identity manifests and is
reinforced, as subjectivity is extracted/imposed by the electronic other. How can such a perception not conjure a
sympathetic response? Yet in that same instant of unity
comes the burning feeling of separation born of envy. The
identity of the electronic body is not our own. We must
eternally consume something to make our appearance more

Body without Organs (first manifestation)

A series of appropriated images appear on 3 TV monitors
which refer to the particular vectors that mark the BwO. As
the images flow across the screens, a silent "body" moves
through the spectators, while 2 voices enunciate the necessity of bodily aphanisis—BwO.

Voice 1: No more cocks. No more cunts. BwO now. All
extensions must be cut off. All orifices must be sewn up—
plugged up. We must rid ourselves of the biological, empty
ourselves of it. All bio-fascism must be ripped out and sealed
up in the clear jars of the museum, so that we will never
forget the pain of somatic tyranny.

Voice 2: For the biggest lie ever was to frame humans as an
organism of consuming, assimilating, incubating, excreting,
creating a whole hierarchy of latent functions.

Voice 1: So we will never forget the late-capitalist physiol-
ogy that bites, sucks, devours—it is driven by the bio-destiny

like its appearance. The desire for greater access to the signs of beauty, health, and intelligence, through the unceasing accumulation of cultural artifacts, brutally reminds us that the perfect excess of the electronic body is not our own. The limitations of the organic abound, and what is achieved becomes vulgar and unnecessary at the point of achievement. All that remains is the unbearable moment of enriched privation. Sympathy and envy are forever spliced together in the form of a hideous Siamese twin. This is the performance of everyday life, so near, so instantaneous, eternally recurring.

---

of the oral hole: consumption, assimilation, incorporation—the mouth must be suppressed, repressed. BwO now.

Voice 2: For too long we have being caught in the circle of the organism, between the goat's anus and the mouth of God, between the logic of the cock and the cunt, the One and the Zero, the cause and the effect—let nothing flow—let nothing pass—BwO now.

Voice 1: The excretion of surplus-value imprisons us in shit-economics: the bio-machine eats in Africa, digests in Asia, and dumps its excess in the first world. The anal force must be eradicated, eliminated. BwO now.

Voice 2: Let us empty the body of its retensions, of its expulsions, of its paranoid dichotomies, of its compulsive production, of its hysterical dissemination, of its neurotic interpretations—let us go further still; we haven't sufficiently dismantled our selves.

Artaud's only misjudgment was his belief that the body without
organs had yet to be created. The electronic body *is* the body
without organs. It already dominates performance, and has
recentered the theater around empty identity and empty
desire. The body without organs is the perfect body—
forever reproducible. No reduction to biology now. Two
hundred Elvis clones appear on the screen. Separate them:
Turn the channel; play the tape. Each performance is on an
eternal loop. These clones were not made in a test tube; they
reproduce of their own accord, each as precise and as perfect
as the last. No fluids, no plagues, no interruptions. The

The "body" kneels before a chair and takes out the "imaginary phallus" and begins to cut it off.

Voice 1: Let us strip ourselves of one part of the body-despot:
an eye, an ear, any piece of epidermis, cut off the cock, sew up
the cunt, plug up the asshole—staple your mouth shut and
remain silent forever. Let us all empty the body.

Voice 2: Let us all empty the body, that coagulated nothing-
ness, and flush it down the toilet: no more shit-economics,
no more urinal-politics.

Voice 2: Let us vanish into the post-biological continuum.

The "body" places the "imaginary phallus" in a clear jar and
seals it, then walks away, leaving the monitors behind.

Voice 2: Dialectical evolution is over—BwO now.

Voices 1 & 2:   BwO now.

orifices of the body without organs are sewn tightly shut. No consumption, no excretion, no interruptions. Such freedom: Safely screened off from the virtual catastrophes of war, capital, gender, or any other manifestation teetering at the brink of a crash, the body without organs is free to drift in the electronic rhizome. The theater of the street and its associated cultural debris collapses. Civilization has been washed clean—progress is complete—dirt, trash, rot, and rubble have been screened off and erased from the perfect world of the electronic body. The electronic body, free of the flesh, free of the economy of desire, has escaped the pain of becoming.

What is the fate of the organic body, caught between sympathy and envy, forever following in the shadow of the body without organs? Very simply, the flesh is sacrificed—carved into layers that better serve various economies. This is not the Cartesian dualism valued by cyberpunk ("Hence, at least through the instrumentality of the Virtual power, mind can exist apart from body, and body apart from mind"), in which the body is no more than a slab of meat. It is not simply a matter of downloading the mind and trashing the body. Rather, the body is divided between surface and depth, between dry and wet. Since spectacle is a dry surface image, the body must reflect that image. The body becomes its mirror, or perhaps more accurately, its xerox. It is paper onto which designer gender, ethnicity, and lifestyle are inscribed. As with any surface of inscription, it must be dry if it is to run through the sight machine. It must also be flat and void of depth (desire). The only acceptable desire is the desire to consume the spectacle's texts. As image cascades down through the various classes of consumption, the resolution of the original decays, until nothing is left but the body as receptacle of water. This is the body sacrificed to the anti-economy. It is the abject body, left to wander the street in misery ("What is sacred undoubtedly corresponds to the object of horror I have spoken of, a fetid, sticky object without boundaries, which teems with life and yet is the sign of death").

The body which signifies the absence of rationalized economic desire is that which we are taught to fear. It is the sign of the organic itself; it is the primordial soup, the placenta-filled womb to which there can be no return. To mention the sacred, or worse, to display signs of the organic, the code of death, is to reject economic inscription. To do so is to become one of the abject, and to suffer great punishment. Many performers have tried to reinstate the organic within the network of value, but they are unable to overcome the power of the body without organs (BwO). The BwO is always there with them, on the stage and in the audience. The best result produced from such work is a cheer for deviance, but the sign of deviance is never broken. Simply putting on a counterspectacle within the theater of the abject is not enough. It only serves to confirm what is already known: Do not mention the organic and its untamed desire, or its yearning for death. Such spectacle is quickly reduced to an aberration, or a peculiar idiosyncrasy. The organic and the electronic must explicitly clash in an attempt to open the rigid hierarchical closure that is presented every day by the engines of the spectacle. To take the most obvious example, this closure is crucial to the success of any horror movie. In every case, horror films express the BwO overcoming the sign of the organic. Spilled guts, sticky goo, splitting skin, erupting pus, uncontrolled excrement, all incite horror in the viewer. It reminds h/er of the organic, that uncontrolled watery excess simply waiting to burst through the seamless xerox surface. The horror movie makes the organic—as well as the means by which it must be punished for its appearance—visible. There are two fundamental rules for simulating horror in spectacular society: The innocent (BwO) must suffer (eat the sacrifice), and the guilty (subelectronic desire) must be punished. The replaying of these two fundamental myths in spectacular endeavors keeps people buying. It makes known that all must aspire to be the innocent and virginal BwO, and that all must block the organic with accumulated piles of manufactured excess. This is the performance that must be disturbed, but it must be disturbed electronically.

If the BwO is conceived of as appearance of self contained in screenal space, it is nearly supernatural to think that the BwO can possess the flesh and walk the earth. It is during the time of possession that the BwO is the most vulnerable to the appearance of organic deficiencies, and yet, this is also the time when the BwO can present itself as an entity separate from spectacle, thus reinforcing its ideal image as existing in the realm of real achievement. The phenomenon of flesh possession by the BwO is commonly referred to as a celebrity. The celebrity acts as empirical proof positive that electronic appearance is but a record of the natural world, and that the electronic is still dependent on the organic. In this form the BwO is not just a mediated screenal vision, but can also be touched, so that it deflects thought away from the categories of the recombinant, and toward the nostalgia of essentialism. Is it any wonder that celebrities are hounded for autographs or any other artifact that can act as a trace of comfort to those desiring the assurances of the pre-electronic order?

The construction of the electronic theater has been completed by
nomadic power. The Situationists alarmed us to its construction when they presented their critique of the
spectacle. Indeed, the melding of architecture, graphic
design, radio, television and film have come to constitute
the spectacular stage, but its logistical support in backstage
virtual technology had yet to fully appear. The strategic
error came when anachronistic forms of resistance (occupa-
tions, strikes, protests, etc.) were used as a means to stop
construction. One of the many failures of the revolutionary
actions of the late 60s and early 70s is that they neither

Body without Organs (second manifestation)

BwO NOW.
BwO NOW.
BwO NOW.

Imperfect flesh is the foundation of screenal economy. The
frenzy of the screenal sign oscillates between perfection and
excess, production and counter-production, panic and hys-
teria. Screenal space inscribes the flesh as the abject. The
screenal space seduces the flesh into the abyss of the surface.
The electronic body is the perfect body. The electronic body
is the body without organs positioned in its screenal space.
It is both self and mirrored self. The electronic body is the
complete body. The body without organs does not decay.
The electronic body does not need the plastic surgeon's
scalpel, liposuction, make-up, or deodorant. It is a body
without organs which cannot suffer, not physiologically,
not psychologically, not sociologically; it is not conscious of
separation. The electronic body seduces those who see it

attacked the electronic theater nor employed nomadic oppositional tactics. The theater of operations was perceived as purely sedentary, without a nomadic component, and was thereby situated in the binary of offense/defense. Within the electronic theater, strategy consists of pure offense. Surveillance systems are the only remaining defensive trace. The trick is never to be caught off guard, always to track the opposition's movements, thus preventing the disappearance of the opponents. The other option is to establish temporary blockage points that allow time to regroup and begin a counter-offensive. The defensive pos-

---

into the bliss of counter-production by offering the hope of a bodily unity that transcends consumption. But the poor, pathetic, organic body is always in a state of becoming. If it consumed just one more product, perhaps it might become whole, perhaps it too could become a body without organs existing in electronic space.

The electronic body oscillates between panic perfection and hysterical aphanisis. The electronic body inscribes the flesh as the abject. At any moment the organic body could fracture and its surface could decay with sickness, ooze and squirt anti-social fluids. The electronic body has shown *ad nauseam* that the spilling of guts, the projecting of vomit, the splitting of skin, the eruption of pus, or any sign of the organic in screenal space exists there only to instill fear, contempt, and embarrassment.

BwO dreams of a body that never existed.
BwO dreams of a body that never existed.
BwO dreams of a body that never existed.
BwO NOW.

ture of fortification is unrealistic. Unfortunately this has traditionally been the tactic (occupation) chosen by the resistance. This was a proper means of resistance against spectacular architecture, but the electronic theater remained untouched and continued expanding its domain. Once again, the culture of resistance is working primarily from a model of critique, and as always, is moving very slowly off the mark in this endeavor, preferring to continue engaging cultural and political bunkers. However, all is not lost. Because of the lack of fortifications in the electronic theater, there are always windows and gaps ripe for disturbance. Unfortunately, such resistance can only come from the technocratic class, and it must occur before surveillance systems become too well-distributed. The performance of the politicized hacker should be the ultimate in performative resistance.

Compared to cyberspace resistance techniques, possible strategies for the cultural producer are much more modest. These producers can re-present the electronic theater for what it is, by creating simulations of performative control that call attention to the technology and methods of control. The other strategy is to attempt to reestablish the organic body in arenas other than the abject and the deviant; however, this performance has no meaning other than to replay the past, unless it is contrasted with the mythic standing of the BwO. To take this approach is not to uncover the invisible, but to impose the vacuum of scepticism on the visible. With either option, the performer must appropriate and occupy the electronic theater. It is unwise to wait until virtual reality has the trappings of a classical theater—one into which the performer and viewer may physically enter and which is enveloped by artificial (electronic) surroundings. As stated earlier, resistant performers must establish those interlocking recombinant stages which oscillate between the theater of everyday life and the virtual theater. Such action will help develop practical performance models—ones which lend themselves to an autonomous performative matrix, rather than ones in which the performers are automatons, replaying the creations of designer culture. Resistant theater is electronic theater.

XI
1776

The mind is a kind of theater. . .There is
properly no simplicity in it at one time, nor
identity in different, only a perpetual flux and
movement, a constant variation, in which
several perceptions successively make their
appearance; pass, re-pass, glide away, and
mingle in an infinite variety of postures and
situations.

The mind is a kind of fractal. . .There is properly
no simplicity in it at one time, nor identity in
different, only a perpetual flux and movement, a
constant variation, in which several perceptions
successively make their appearance; pass, re-
pass, glide away, and mingle in an infinite variety
of DNA and recombinations.

XII
1819

It then becomes clear and certain to him that
what he knows is not a sun and an earth, but
only an eye that sees a sun, a hand that feels
an earth; that the world which surrounds him
is there only as idea.

It then becomes clear and certain to him that
what he knows is not a sun and an earth, but only
an eye visor that sees a sun, a data glove that
feels an earth; that the integrated world which
surrounds him is there only as simulacra.

Recombinant text, circa 1500.

5

Utopian Plagiarism,

Hypertextuality, and

Electronic Cultural Production

Plagiarism has long been considered an evil in the cultural world.
Typically it has been viewed as the theft of language, ideas,
and images by the less than talented, often for the enhance-
ment of personal fortune or prestige. Yet, like most
mythologies, the myth of plagiarism is easily inverted.
Perhaps it is those who support the legislation of representation and the privatization of language that are suspect;
perhaps the plagiarist's actions, given a specific set of social
conditions, are the ones contributing most to cultural en-
richment. Prior to the Enlightenment, plagiarism was useful
in aiding the distribution of ideas. An English poet could
appropriate and translate a sonnet from Petrarch and call it

A version of this article was originally published in *Critical Issues in Electronic Media*.
Simon Penny, ed. New York: SUNY Press, 1994.

his own. In accordance with the classical aesthetic of art as imitation, this was a perfectly acceptable practice. The real value of this activity rested less in the reinforcement of classical aesthetics than in the distribution of work to areas where otherwise it probably would not have appeared. The works of English plagiarists, such as Chaucer, Shakespeare, Spenser, Sterne, Coleridge, and De Quincey, are still a vital part of the English heritage, and remain in the literary canon to this day.

At present, new conditions have emerged that once again make plagiarism an acceptable, even crucial strategy for textual production. This is the age of the recombinant: recombinant bodies, recombinant gender, recombinant texts, recombinant culture. Looking back through the privileged frame of hindsight, one can argue that the recombinant has always been key in the development of meaning and invention; recent extraordinary advances in electronic technology have called attention to the recombinant both in theory and in practice (for example, the use of morphing in video and film). The primary value of all electronic technology, especially computers and imaging systems, is the startling speed at which they can transmit information in both raw and refined forms. As information flows at a high velocity through the electronic networks, disparate and sometimes incommensurable systems of meaning intersect, with both enlightening and inventive consequences. In a society dominated by a "knowledge" explosion, exploring the possibilities of meaning in that which already exists is more pressing than adding redundant information (even if it is produced using the methodology and metaphysic of the "original"). In the past, arguments in favor of plagiarism were limited to showing its use in resisting the privatization of culture that serves the needs and desires of the power elite. Today one can argue that plagiarism is acceptable, even inevitable, given the nature of postmodern existence with its techno-infrastructure. In a recombinant culture, plagiarism is productive, although we need not abandon the romantic model of cultural production which privileges a model of *ex nihilo* creation. Certainly in a general sense the latter model is somewhat anachronistic. There are still specific situations where such thinking is useful, and one can never be sure when it could become appropriate again. What is called for is an end to its tyranny and to its institutionalized cultural bigotry. This is a call to open the cultural data base, to let everyone use the technology of textual production to its maximum potential.

> Ideas improve. The meaning of words participates in the improvement. Plagiarism is necessary. Progress implies it. It embraces an author's phrase, makes use of his expressions, erases a false idea, and replaces it with the right idea.[1]

Plagiarism often carries a weight of negative connotations (particularly in the bureaucratic class); while the need for its use has increased over the century, plagiarism itself has been camouflaged in a new lexicon by those desiring to explore the practice as method and as a legitimized form of cultural discourse. Readymades, collage, found art or found text, intertexts, combines, detournment, and appropriation—all these terms represent explorations in plagiarism. Indeed, these terms are not perfectly synonymous, but they all intersect a set of meanings primary to the philosophy and activity of plagiarism. Philosophically, they all stand in opposition to essentialist doctrines of the text: They all assume that no structure within a given text provides a universal and necessary meaning. No work of art or philosophy exhausts itself in itself alone, in its being-in-itself. Such works have always stood in relation to the actual life-process of society from which they have distinguished themselves. Enlightenment essentialism failed to provide a unit of analysis that could act as a basis of meaning. Just as the connection between a signifier and its referent is arbitrary, the unit of meaning used for any given textual analysis is also arbitrary. Roland Barthes' notion of the lexia primarily indicates surrender in the search for a basic unit of meaning. Since language was the only tool available for the development of metalanguage, such a project was doomed from its inception. It was much like trying to eat soup with soup. The text itself is fluid—although the language game of ideology can provide the illusion of stability, creating blockage by manipulating the unacknowledged assumptions of everyday life. Consequently, one of the main goals of the plagiarist is to restore the dynamic and unstable drift of meaning, by appropriating and recombining fragments of culture. In this way, meanings can be produced that were not previously associated with an object or a given set of objects.

Marcel Duchamp, one of the first to understand the power of recombination, presented an early incarnation of this new aesthetic with his readymade series. Duchamp took objects to which he was "visually indifferent," and recontextualized them in a manner that shifted their meaning. For example, by taking a urinal out of the rest room, signing it, and placing it on a pedestal in an art gallery, meaning slid away from the apparently exhaustive functional interpretation of the object. Although this meaning did not completely disappear, it was placed in harsh juxta-position to another possibility—meaning as an art object. This problem of instability increased when problems of origin were raised: The object was not made by an artist, but by a machine. Whether or not the viewer chose to accept other possibilities for interpreting the function of the artist and the authenticity of the art object, the urinal in a gallery instigated a moment of uncertainty and reassessment. This conceptual game has been replayed numerous times over the 20th century, at times for very narrow purposes, as with Rauschenberg's combines—done for the sake of attacking the critical hegemony of Clement Greenberg—while at other times it has been done to promote large-scale political and cultural restructuring, as in the case of the Situationists. In each case, the plagiarist works to open meaning through the injection of scepticism into the culture-text.

Here one also sees the failure of Romantic essentialism. Even the alleged transcendental object cannot escape the sceptics' critique. Duchamp's notion of the inverted readymade (turning a Rembrandt painting into an ironing board) suggested that the distinguished art object draws its power from a historical legitimation process firmly rooted in the institutions of western culture, and not from being an unalterable conduit to transcendental realms. This is not to deny the possibility of transcendental experience, but only to say that if it does exist, it is prelinguistic, and thereby relegated to the privacy of an individual's subjectivity. A society with a complex division of labor requires a rational-ization of institutional processes, a situation which in turn robs the individual of a way to share nonrational experience. Unlike societies with a simple division of labor, in which the experience of one member closely resembles the experience of another (minimal alienation), under a complex division of labor, the life experience of the individual turned specialist holds little in common with other specialists. Consequently, communication exists primarily as an instrumental function.

Plagiarism has historically stood against the privileging of any text through spiritual, scientific, or other legitimizing myths. The plagiarist sees all objects as equal, and thereby horizontalizes the plane of phenomena. All texts become potentially usable and reusable. Herein lies an epistemology of anarchy, according to which the plagiarist argues that if science, religion, or any other social institution precludes certainty beyond the realm of the private, then it is best to endow consciousness with as many categories of interpretation as possible. The tyranny of paradigms may have some useful consequences (such as greater efficiency within the paradigm), but the repressive costs to the individual (excluding other modes of thinking and reducing the possibility of invention) are too high. Rather than being led by sequences of signs, one should instead drift through them, choosing the interpretation best suited to the social conditions of a given situation.

> It is a matter of throwing together various cut-up techniques in order to respond to the omnipresence of transmitters feeding us with their dead discourses (mass media, publicity, etc.). It is a question of unchaining the codes—not the subject anymore—so that something will burst out, will escape; words beneath words, personal obsessions. Another kind of word is born which escapes from the totalitarianism of the media but retains their power, and turns it against their old masters.

Cultural production, literary or otherwise, has traditionally been a slow, labor-intensive process. In painting, sculpture, or written work, the technology has always been primitive by contemporary standards. Paintbrushes, hammers and chisels, quills and paper, and even the printing press do not lend themselves well to rapid production and broad-range distribution. The time lapse between production and distribution can seem unbearably long. Book arts and traditional visual arts still suffer tremendously from this problem, when compared to the electronic arts. Before electronic technology became dominant, cultural perspectives developed in a manner that more clearly defined texts as individual works. Cultural fragments appeared in their own right as discrete units, since their influence moved slowly enough to allow the orderly evolution of an argument or an aesthetic. Boundaries could be maintained between disciplines and schools of thought. Knowledge was considered finite, and was therefore easier to control. In the 19th century this traditional order began to collapse as new technology began to increase the velocity of cultural development. The first strong indicators began to appear that speed was becoming a crucial issue. Knowledge was shifting away from certitude, and transforming itself into information. During the American Civil War, Lincoln sat impatiently by his telegraph line, awaiting reports from his generals at the front. He had no patience with the long-winded rhetoric of the past, and demanded from his generals an efficient economy of language. There was no time for the traditional trappings of the elegant essayist. Cultural velocity and information have continued to increase at a geometric rate since then, resulting in an information panic. Production and distribution of information (or any other product) must be immediate; there can be no lag time between the two. Techno-culture has met this demand with data bases and electronic networks that rapidly move any type of information.

Under such conditions, plagiarism fulfills the requirements of economy of representation, without stifling invention. If invention occurs when a new perception or idea is brought out—by intersecting two or more formally disparate systems—then recombinant methodologies are desirable. This is where plagiarism progresses beyond nihilism. It does not simply inject scepticism to help destroy totalitarian systems that stop invention; it participates in invention, and is thereby also productive. The genius of an inventor like Leonardo da Vinci lay in his ability to recombine the then separate systems of biology, mathematics, engineering, and art. He was not so much an originator as a synthesizer. There have been few people like him over the centuries, because the ability to hold that much data in one's own biological memory is rare. Now, however, the technology of recombination is available in the computer. The problem now for would-be cultural producers is to gain access to this technology and information. After all, access is the most precious of all privileges, and is therefore strictly guarded, which in turn makes one wonder whether to be a successful plagiarist, one must also be a successful hacker.

> Most serious writers refuse to make themselves available to the things that technology is doing. I have never been able to understand this sort of fear. Many are afraid of using tape recorders, and the idea of using any electronic means for literary or artistic purposes seems to them some sort of sacrilege.

To some degree, a small portion of technology has fallen through the cracks into the hands of the lucky few. Personal computers and video cameras are the best examples. To accompany these consumer items and make their use more versatile, hypertextual and image sampling programs have also been developed—programs designed to facilitate recombination. It is the plagiarist's dream to be able to call up, move, and recombine text with simple user-friendly commands. Perhaps plagiarism rightfully belongs to post-book culture, since only in that society can it be made explicit what book culture, with its geniuses and auteurs, tends to hide—that information is most useful when it interacts with other information, rather than when it is deified and presented in a vacuum.

Thinking about a new means for recombining information has always been on 20th-century minds, although this search has been left to a few until recently. In 1945 Vannevar Bush, a former science advisor to Franklin D. Roosevelt, proposed a new way of organizing information in an *Atlantic Monthly* article. At that time, computer technology was in its earliest stages of development and its full potential was not really understood. Bush, however, had the foresight to imagine a device he called the Memex. In his view it would be based around storage of information on microfilm, integrated with some means to allow the user to select and display any section at will, thus enabling one to move freely among previously unrelated increments of information.

At the time, Bush's Memex could not be built, but as computer technology evolved, his idea eventually gained practicality. Around 1960 Theodor Nelson made this realization when he began studying computer programming in college:

Over a period of months, I came to realize that, although programmers structured their data hierarchically, they didn't have to. I began to see the computer as the ideal place for making interconnections among things accessible to people.

I realized that writing did not have to be sequential and that not only would tomorrow's books and magazines be on [cathode ray terminal] screens, they could all tie to one another in every direction. At once I began working on a program (written in 7090 assembler language) to carry out these ideas.

Nelson's idea, which he called hypertext, failed to attract any supporters at first, although by 1968 its usefulness became obvious to some in the government and in defense industries. A prototype of hypertext was developed by another computer innovator, Douglas Englebart, who is often credited with many breakthroughs in the use of computers (such as the development of the Macintosh interface, Windows). Englebart's system, called Augment, was applied to organizing the government's research network, ARPAnet, and was also used by McDonnell Douglas, the defense contractor, to aid technical work groups in coordinating projects such as aircraft design:

All communications are automatically added to the Augment information base and linked, when appropriate, to other documents. An engineer could, for example, use Augment to write and deliver electronically a work plan to others in the work group. The other members could then review the document and have their comments linked to the original, eventually creating a "group memory" of the decisions made. Augment's powerful linking features allow users to find even old information quickly, without getting lost or being overwhelmed by detail.

Computer technology continued to be refined, and eventually—as with so many other technological breakthroughs in this country—once it had been thoroughly exploited by military and intelligence agencies, the technology was released for commercial exploitation. Of course, the development of microcomputers and consumer-grade technology for personal computers led immediately to the need for software which would help one cope with the exponential increase in information, especially textual information. Probably the first humanistic application of hypertext was in the field of education. Currently, hypertext and hypermedia (which adds graphic images to the network of features which can be interconnected) continue to be fixtures in instructional design and educational technology.

An interesting experiment in this regard was instigated in 1975 by Robert Scholes and Andries Van Dam at Brown University. Scholes, a professor of English, was contacted by Van Dam, a professor of computer science, who wanted to know if there were any courses in the humanities that might benefit from using what at the time was called a text-editing system (now known as a word processor) with hypertext capabilities built in. Scholes and two teaching assistants, who formed a research group, were particularly impressed by one aspect of hypertext. Using this program would make it possible to peruse in a nonlinear fashion all the interrelated materials in a text. A hypertext is thus best seen as a web of interconnected materials. This description suggested that there is a definite parallel between the conception of culture-text and that of hypertext:

> One of the most important facets of literature (and one which also leads to difficulties in interpretation) is its reflexive nature. Individual poems constantly develop their meanings—often through such means as direct allusion or the reworking of traditional motifs and conventions, at other times through subtler means, such as genre development and expansion or biographical reference—by referring to that total body of poetic material of which the particular poems comprise a small segment.

Although it was not difficult to accumulate a hypertextually-linked data base consisting of poetic materials, Scholes and his group were more concerned with making it interactive—that is, they wanted to construct a "communal text" including not only the poetry, but also incorporating the comments and interpretations offered by individual students. In this way, each student in turn could read a work and attach "notes" to it about his or her observations. The resulting "expanded text" would be read and augmented at a terminal on which the screen was divided into four areas. The student could call up the poem in one of the areas (referred to as windows) and call up related materials in the other three windows, in any sequence he or she desired. This would powerfully reinforce the tendency to read in a nonlinear sequence. By this means, each student would learn how to read a work as it truly exists, not in "a vacuum" but rather as the central point of a progressively-revealed body of documents and ideas.

Hypertext is analogous to other forms of literary discourse besides poetry. From the very beginning of its manifestation as a computer program, hypertext was popularly described as a multidimensional text roughly analogous to the standard scholarly article in the humanities or social sciences, because it uses the same conceptual devices, such as footnotes, annotations, allusions to other works, quotations from other works, etc. Unfortunately, the convention of linear reading and writing, as well as the physical fact of two-dimensional pages and the necessity of binding them in only one possible sequence, have always limited the true potential of this type of text. One problem is that the reader is often forced to search through the text (or forced to leave the book and search elsewhere) for related information. This is a time-consuming and distracting process; instead of being able to move easily and instantly among physically remote or inaccessible areas of information storage, the reader must cope with cumbrous physical impediments to his or her research or creative work. With the advent of hypertext, it has become possible to move among related areas of information with a speed and flexibility that at least approach finally accommodating the workings of human intellect, to a degree that books and sequential reading cannot possibly allow.

> The recombinant text in hypertextual form signifies the emergence of the perception of textual constellations that have always/already gone nova. It is in this uncanny luminosity that the authorial biomorph has been consumed.[2]

Barthes and Foucault may be lauded for theorizing the death of the author; the absent author is more a matter of everyday life, however, for the technocrat recombining and augmenting information at the computer or at a video editing console. S/he is living the dream of capitalism that is still being refined in the area of manufacture. The Japanese notion of "just in time delivery," in which the units of assembly are delivered to the assembly line just as they are called for, was a first step in streamlining the tasks of assembly. In such a system, there is no sedentary capital, but a constant flow of raw commodities. The assembled commodity is delivered to the distributor precisely at the moment of consumer need. This nomadic system eliminates stockpiles of goods. (There still is some dead time; however, the Japanese have cut it to a matter of hours, and are working on reducing it to a matter of minutes). In this way, production, distribution, and consumption are imploded into a single act, with no beginning or end, just unbroken circulation. In the same manner, the online text flows in an unbroken stream through the electronic network. There can be no place for gaps that mark discrete units in the society of speed. Consequently, notions of origin have no place in electronic reality. The production of the text presupposes its immediate distribution, consumption, and revision. All who participate in the network also participate in the interpretation and mutation of the textual stream. The concept of the author did not so much die as it simply ceased to function. The author has become an abstract aggregate that cannot be reduced to biology or to the psychology of personality. Indeed, such a development has apocalyptic connotations—the fear that humanity will be lost in the textual stream. Perhaps humans are not capable of participating in hypervelocity. One must answer that never has there been a time when humans were able, one and all, to participate in cultural production. Now, at least the potential for cultural democracy is greater. The single bio-genius need not act as a stand-in for all humanity. The real concern is just the same as it has always been: the need for access to cultural resources.

> The discoveries of postmodern art and criticism regarding the analogical structures of images demonstrate that when two objects are brought together, no matter how far apart their contexts may be, a relationship is formed. Restricting oneself to a personal relationship of words is mere convention. The bringing together of two independent expressions supersedes the original elements and produces a synthetic organization of greater possibility.[3]

The book has by no means disappeared. The publishing industry continues to resist the emergence of the recombinant text, and opposes increases in cultural speed. It has set itself in the gap between production and consumption of texts, which for purposes of survival it is bound to maintain. If speed is allowed to increase, the book is doomed to perish, along with its renaissance companions painting and sculpture. This is why the industry is so afraid of the recombinant text. Such a work closes the gap between production and consumption, and opens the industry to those other than the literary celebrity. If the industry is unable to differentiate its product through the spectacle of originality and uniqueness, its profitability collapses. Consequently, the industry plods along, taking years to publish information needed immediately. Yet there is a peculiar irony to this situation. In order to reduce speed, it must also participate in velocity in its most intense form, that of spectacle. It must claim to defend "quality and standards," and it must invent celebrities. Such endeavors require the immediacy of advertising—that is, full participation in the simulacra that will be the industry's own destruction.

Hence for the bureaucrat, from an everyday life perspective, the author is alive and well. S/he can be seen and touched and traces of h/is existence are on the covers of books and magazines everywhere in the form of the signature. To such evidence, theory can only respond with the maxim that the meaning of a given text derives exclusively from its relation to other texts. Such texts are contingent upon what came before them, the context in which they are placed, and the interpretive ability of the reader. This argument is of course unconvincing to the social segments caught in cultural lag. So long as this is the case, no recognized historical legitimation will support the producers of recombinant texts, who will always be suspect to the keepers of "high" culture.

> Take your own words or the words said to be "the very own words" of anyone else living or dead. You will soon see that words do not belong to anyone. Words have a vitality of their own. Poets are supposed to liberate the words—not to chain them in phrases. Poets have no words "of their very own." Writers do not own their words. Since when do words belong to anybody? "Your very own words" indeed! and who are "you"?

The invention of the video portapak in the late 1960s and early 70s led to considerable speculation among radical media artists that in the near future, everyone would have access to such equipment, causing a revolution in the television industry. Many hoped that video would become the ultimate tool for distributable democratic art. Each home would become its own production center, and the reliance on network television for electronic information would be only one of many options. Unfortunately this prophecy never came to pass. In the democratic sense, video did little more than super 8 film to redistribute the possibility for image production, and it has had little or no effect on image distribution. Any video besides home movies has remained in the hands of an elite technocratic class, although (as with any class) there are marginalized segments which resist the media industry, and maintain a program of decentralization.

The video revolution failed for two reasons—a lack of access and an absence of desire. Gaining access to the hardware, particularly post-production equipment, has remained as difficult as ever, nor are there any regular distribution points beyond the local public access offered by some cable TV franchises. It has also been hard to convince those outside of the technocratic class why they should want to do something with video, even if they had access to equipment. This is quite understandable when one considers that media images are provided in such an overwhelming quantity that the thought of producing more is empty. The contemporary plagiarist faces precisely the same discouragement. The potential for generating recombinant texts at present is just that, potential. It does at least have a wider base, since the computer technology for making recombinant texts has escaped the technocratic class and spread to the bureaucratic class; however, electronic cultural production has by no means become the democratic form that utopian plagiarists hope it will be.

The immediate problems are obvious. The cost of technology for productive plagiarism is still too high. Even if one chooses to use the less efficient form of a hand-written plagiarist manuscript, desktop publishing technology is required to distribute it, since no publishing house will accept it. Further, the population in the US is generally skilled only as receivers of information, not as producers. With this exclusive structure solidified, technology and the desire and ability to use it remain centered in utilitarian economy, and hence not much time is given to the technology's aesthetic or resistant possibilities.

In addition to these obvious barriers, there is a more insidious problem that emerges from the social schizophrenia of the US. While its political system is theoretically based on democratic principles of inclusion, its economic system is based on the principle of exclusion. Consequently, as a luxury itself, the cultural superstructure tends towards exclusion as well. This economic principle determined the invention of copyright, which originally developed not in order to protect writers, but to reduce competition among publishers. In 17th-century England, where copyright first appeared, the goal was to reserve for publishers themselves, in perpetuity, the exclusive right to print certain books. The justification, of course, was that when formed into a literary work, language has the author's personality imposed upon it, thereby marking it as private property. Under this mythology, copyright has flourished in late capital, setting the legal precedent to privatize any cultural item, whether it is an image, a word, or a sound. Thus the plagiarist (even of the technocratic class) is kept in a deeply marginal position, regardless of the inventive and efficient uses h/is methodology may have for the current state of technology and knowledge.

> What is the point of saving language when there is
> no longer anything to say?

The present requires us to rethink and re-present the notion of plagiarism. Its function has for too long been devalued by an ideology with little place in techno-culture. Let the romantic notions of originality, genius, and authorship remain, but as elements for cultural production without special privilege above other equally useful elements. It is time to openly and boldly use the methodology of recombination so as to better parallel the technology of our time.

Notes

1   In its more heroic form the footnote has a low-speed hypertextual function—that is, connecting the reader with other sources of information that can further articulate the producer's words. It points to additional information too lengthy to include in the text itself. This is not an objectionable function. The footnote is also a means of surveillance by which one can "check up" on a writer, to be sure that s/he is not improperly using an idea or phrase from the work of another. This function makes the footnote problematic, although it may be appropriate as a means of verifying conclusions in a quantitative study, for example. The surveillance function of the footnote imposes fixed interpretations on a linguistic sequence, and implies ownership of language and ideas by the individual cited. The note becomes an homage to the genius who supposedly originated the idea. This would be acceptable if all who deserved credit got their due; however, such crediting is impossible, since it would begin an infinite regress. Consequently, that which is most feared occurs: the labor of many is stolen, smuggled in under the authority of the signature which is cited. In the case of those cited who are still living, this designation of authorial ownership allows them to collect rewards for the work of others. It must be realized that writing itself is theft: it is a changing of the features of the old culture-text in much the same way one disguises stolen goods. This is not to say that signatures should never be cited; but remember that the signature is merely a sign, a shorthand under which a collection of interrelated ideas may be stored and rapidly deployed.

2   If the signature is a form of cultural shorthand, then it is not necessarily horrific on occasion to sabotage the structures so they do not fall into rigid complacency. Attributing words to an image, i.e., an intellectual celebrity, is inappropriate. The image is a tool for playful use, like any culture-text or part thereof. It is just as necessary to imagine the history of the spectacular image, and write it as imagined, as it is to show fidelity to its current "factual" structure. One should choose the method that best suits the context of production, one that will render the greater possibility for interpretation. The producer of recombinant texts augments the language, and often preserves the generalized code, as when Karen Eliot quoted Sherrie Levine as saying, "Plagiarism? I just don't like the way it tastes."

3   It goes without saying that one is not limited to correcting a work or to integrating diverse fragments of out-of-date works into a new one; one can also alter the meaning of these fragments in any appropriate way, leaving the constipated to their slavish preservation of "citations."

Four examples of plagiarist poetry.

Like A Big Dog*

A big dog stands on the highway
He walks on confidently and is run over by a car.
His peaceful expression shows that he is usually better looked after—
a domestic animal to whom no harm is done. * *
But do the sons of the rich bourgeois families
who also suffer no harm* * *
have the same peaceful expression?
They were cared for just as lovingly
as the dog which is now run over.

Annotations for Like a Big Dog

* From Horkheimer & Adorno, *Dialectic of Enlightenment*, "Animal Psychology."

* * In Kafka's "Investigations of a Dog" the same dog is referred to as "impossible to abuse and impossible to love."

* * * a reversal of the German expression "the wealthy fear harm for they cause most of it."

Crônicas III

The one who told me the story was a very dear friend.
The child was a little Indian boy, really quite small.
All the members of the tribe took care of the manioc patch.
The new buildings were very daring constructions.*
He expected the child to have a shock when he saw all those
    apartments in just one building.
However, the sight had no effect except for a yawn.
"When are we going to visit the theaters, the banks, and the
    squares?" he asked with impatience.
To me, your attitude is completely incomprehensible.
The interest we show is related to our own lives.
Without fortune and a good car, our social group feels there
    can be
no well-being.**

* To show local tribes the value of the paper industry that
was destroying the jungle in which they lived, the corpora-
tion built huts made of corrugated cardboard for the
tribespeople.
**The motto of one of the Samba troupes, most of which
come from the poorest sections of Rio and dress like wealthy
aristocrats during Carnival.

Narkotika I

this is the diseasing of America.
Normal joy and pain are denied us,
through being defined as clinical syndromes.

our failure will differ from that of previous civilizations,
in that our demise will be scientific.
Medical treatments will expand endlessly
but will not be able to help us.
In this perverted medical effort, we lose hope.
Disease conceptions have come to stand for all our fears.

While we rush to spend money in new ways,
More seek treatment for the disorder
Only to relapse, and the very failures of
disease treatment are cited as proof of its effectiveness.

One reaction to a dearth of cultural theory

A few theoretical issues in the study of modern systems:
material objects are not part of culture.
certain cultural performances create wastes that
are products, not parts, of the culture proper.
Confining an earthworm, a snail, and a chicken
together in one box does not make them members of the
        same species.
No modern system is completely consistent or compatible.
For example, in our system the manufacture of rubber heels
        for shoes
is in neutral consistency with the professional study of
        literature.
The use of the slang word "shucks" has little or nothing to do
with our system's adjustment to its environment
or with its relations with foreign cultures.
Let us ask again how they can be held together.
The answer that many would give is "force."

XIII
1832

Thou buildest upon the bosom of darkness,
out of the fantastic imagery of the brain, cities
and temples, beyond the art of Phideas and
Praxiteles, beyond the splendors of Babylon
and Hekatómpylos; and, "from the anarchy of
dreaming sleep," callest into sunny light the
faces of long buried beauties.

Thou buildest upon the bosom of darkness, out of
the fantastic imagery of the brain, cities and
temples of digital perfection, beyond the art of
Phideas and Praxiteles, beyond the splendors of
Babylon and Hekatómpylos; and, "from the
anarchy of dreaming sleep," callest into cathode
light the faces of long buried beauties.

XIV
1843

What is abstract thought? It is thought without a thinker. Abstract thought ignores everything except the thought, and only the thought is, and is in its own medium.

What is virtual thought? It is thought without a thinker. Virtual thought ignores everything except the thought, and only the thought is, and is in its own medium.

6

Fragments on the

Problem of Time

Sites and methods of resistance have traditionally been defined in terms of space. The goal of most resistant action has been to destabilize a limited physical space, on the assumption that power, like the society in which it is housed, is sedentary and confined to a fixed geographic location. However, recent technological advances have brought out the need to reassess spatial disturbance as the only productive form of resistance. To be sure, the nature of power itself has fundamentally changed. No longer intimately tied to state space, it has recentered itself in the free zone of time. Power has shed as many of its sedentary attachments as possible, so that where it is located matters less than the speed of its movement between temporary points of blockage, and the time needed to remove blockage. With the emergence of cyber networks, authoritarian space can be folded and carried to any point on the electronic rhizome. The war machine has shifted its strategy away from the centralized fortress to a decentralized, deterritorialized, floating field. It has become disembodied. The ideology which parallels this economic shift has yet to really congeal; the ideology of the sedentary is still dominant. Given this situation, one of the key objectives of the resistant cultural worker is to disturb the solidification of the new ideology before it becomes a symbolic order of even greater tyranny than the current one, and to rechannel the convergence of hardware (video, telephone, and computer) into a decentralized form accessible to others besides the power elite. Before this nearly impossible task can be attempted, cultural workers must step back and use time, rather than space, as a frame for analyzing the priorities of resistance.

This is not a call for a return to historicism, or to any other modernist notion of time, as it is not really possible to differentiate between fiction and history in a period of information overload. The perpetually rolling avalanche of information has not clarified the current situation, but has confused it, leaving the once secure binaries of the dialectic in a state of ruins. History is no more: Only speculative reflection remains on what is now the fractal of time. The greater the speed, the greater the intensity of fragmentation. There are traces of modern thought linked to this discourse, since fragmentation was central to discussions of the complex division of labor, both general and specific, among theorists such as Marx, Weber, Adorno, the Situationists, etc. Yet the division of labor as a historical backbone at the macro level, or as critique of assembly-line oppression at the micro level, now is insufficient to describe and explain separation.

The notion of cultural lag has been a part of sociological discourse from its beginnings, since it has long been suggested that differing sectors of society move at different rates of speed. In the society of late capital, as in most societies, the economy—ever hungry for greater production efficiency and new product development—has traveled the quickest. The supreme economic value of maintaining an edge over competitors by advancing production techniques and distribution tactics, while shortening the duration of research and development, has become impossible to integrate with other value systems. Typically, ideology (state-sanctioned value systems) is just the opposite in its resistance to new values. Changes in ideology are very slow, since in the grandest sense ideology consists of master narratives that offer the illusion of stability and security necessary to make everyday life intelligible. There is a peculiar contradiction between economic and "moral" ideology, since the latter can act as a resister to the former. Many of the biblical master narratives, for instance, are at odds with the value of speed so essential to the economic sector. The explanation for this contradiction is found in the political sector. Its function is to mediate the contradictions. As an arm of the economy, the legitimized political sector has the unenviable task of keeping the economy as free of regulation as possible, while seeming to meet contradictory cultural demands. For example, the master narrative of the welfare state has been a key site of inertia in the United States. The idea that the destitute must be given a second chance, the sick be cared for, and the ignorant be educated, is antithetical to the construction and maintenance of bourgeois economy. The government's role in this conflict is to maintain a symbolic order conducive to the perception that the welfare state is functioning on behalf of its citizens, while allowing the business sector to follow its anti-welfare agenda. This can be done, for example, by suggesting small increases in the minimum wage, while signing free-trade agreements with third world countries which allow unrestricted colonization of their labor pools. Many times, administrations in the US change because there is a crisis in the perception that the demands of the welfare state are being met.

This discussion about the notion of cultural lag points out how different present-day institutions are simultaneously situated in different historical time zones. To complicate matters further, even the components of each institution are not necessarily in the same time zone. The US military is an example of an institution that has advanced the furthest into the future, a world alien to everyday life perception. The capabilities of its technology and the means of its deployment almost defy imagination. Such components are structured by nomadic values, using the idea of globalized control through absence as a master narrative. Yet beyond this one narrative, the ideological component in the military is extremely conflicted. Its interrelationship with the government continually pulls it away from an ideology of globalization to one of the nation-state, and with that retrogression come all the questions about women in combat or the acceptance of gays in the military. In a time of alienated electro-mechanized war, flesh values would seem irrelevant from any perspective. But to admit this explicitly is too disruptive to authoritarian institutions still living under the symbolic order of imperialism; in this time zone racism and sexism still have a necessary function, as they benefit these institutions' exploitative aims (for example, to camouflage the corporate need to maintain a reserve labor force), as well as justifying obscene re-appropriation. Consequently it falls to the government to seek a compromise between the two time fragments.

The fractalization of lived time occurs not only in abstracted macro institutions, but also exists at the micro level of everyday life, as well as at the intermediate level of social aggregates. To be sure, the constructs of race/ethnicity, gender, and class can also be factored in. The life-world of individuals in the technocratic class undoubtedly stands in marked contrast to those in the working class, partly because of extreme differences in production technology. While the former class works in an electronic environment that is constantly transforming, the latter still proceeds according to a model of production that at best has entered the time zone of post-industrial machine technology. In terms of mediated leisure, the two may share a similar time zone, since both have access to television, but this is more a by-product of the market continuum that intersects all time zones. Everyday life itself becomes a determined walk through given segments of history without ever leaving the present.

Much of authoritarian power now functions to control the time zones to which an individual has access, and this is precisely the problem when race and gender are examined. Frustration caused by the inability to solve spatial problems (which in turn are represented by imperialist ideology, such as prejudgment determined by spatial representation, or ghetto lock-downs) is not the only reason that race and gender relations have reached such a point of crisis; there is also the matter of temporal lock-down. Numerous social aggregates are caught in the time zone of imperialism. The colonial era of conquest is constantly replayed, even though the conquest by interdependent economic and military superpowers is complete. There is no more territory to appropriate; it can only be re-appropriated (for example, family farms). Yet areas lacking premium value as market territories or as strategic militarized zones remain in the historical void of imperialism. Sexism and racism no longer act as justifications for expansion, but rather as justifications to mark these territories as sites for sacrificial waste inherent to the capitalist system. Time has stopped for those caught in these territories. The future cannot be accessed, although some narrow conduits out of these areas have been constructed. This is especially true for straight white women, as they have had more middle class support. However, the more that marginals advance toward the future and away from their previous temporal sites, the greater the expectations of those on the move as well as those who are left behind. With these expectations come the realization that full spectrum temporal mobility is highly improbable, thus dramatically increasing frustration and anger. Running parallel to this problem is that of splintering perspective. With each new time zone entered, new theoretical and practical considerations arise. (For example, in the time zone of imperialism, theorizing and implementing communal neighborhood defense systems is a necessity, while in the late capital time zone of cyberspace the theorization and implementation of cell attack strategies has a more viable function). As groups move through time, their perspectives fragment. This is why essentialist and nomadic positions can both seem to be true. The former lags behind the latter, but each has time zones in which it is ascendant. The essentialist position functions best in the time zones of early capital, while the nomadic position functions best in the time zones of late capital.

What should become clear from this discussion is that there is not a monolithic historical present. The present has been shattered into a thousand shards, all of which require different strategies for resistance. Now more than ever, an anarchist epistemology should be adopted, one that leads to situational knowledge. It must be one that tolerates research and exploration within any time or spatial zone. Resistance cannot be carried out from the safety zone of a single bunker. Those who are able must be free to move through time by any means necessary.

The situation of the resistant cultural worker in regard to the problem of a shattered historical present is quite peculiar. Here is a class of workers with relative autonomy in regard to the historical time zone in which they choose to work, and yet they tend to remain solely invested in a reductive resistance to imperialist ideology. From the position of the cultural worker, concerns are usually framed around questions of identity and colonization. There is no doubt that this is a key site of struggle, but too many resources have been deployed in this sector. The degree of redundancy and reinvention occurring in this time zone is unfortunate primarily because it offers a spectacle that implies that other time zones do not exist, or that they are irrelevant, and that no other problem can be solved or tended to until the imperialist mess of early capital is cleaned up. Consequently, authoritarian power is allowed to run unchecked in other time zones, constructing and reconstructing the worker and itself in a manner most beneficial to its concerns. From the perspective of past historical time zones, the idea of class analysis—still an incomplete project—has been severely undermined. This loss has removed a functional category for understanding marginalization beyond that which reduces the world to the appearance of flesh. The recent appropriation of class critique by the Democrats, in an effort to dissolve the radical idea of class warfare under the sign of liberal reformism, demonstrates how much is being surrendered without resistance from the cultural worker in order to perpetuate a discourse of identity that now teeters on the brink of full-scale commodification. On the other end of the time continuum, these concerns about identity and power make explorations into technology and the emergence of electronic space seem unnecessary. It must then be asked, has identity politics become a code of entrapment? Is it a code of liberation or one of tyranny within the realm of cultural production?

One of the most well-rehearsed and routinized performances to arrive as fallout from identity discourse is the call and response ritual which asks, "Who created authoritarian culture?" and "Who benefits from this culture?" The response from the chorus is "The straight white male." Within this discourse and performance matrix, the identity of the straight white male is solely informed by his role as an irredeemable criminal. It is odd to think that whether one's perspective is from the margin or the center, evil is always incarnated in the flesh. This has been the primary failure of identity politics thus far. The Christian master narrative in which evil is reducible to flesh has kept its structure intact. Although the contingent elements of this narrative have been inverted—Eve as the innocent and Adam as the guilty—the original sin of the flesh continues ever onward. Everyone knows who the criminals are since they can easily be recognized; they are forever marked by the fleshy appearance of their genetic code. Such are the wages of sin, and such is the foundation for the ideology of exploitation.

Social solidarity among those of resistant culture cannot be based on the same principle as that of authoritarian culture. To do so is to perpetuate the mechanisms of exclusion and elimination, which in turn preserves the rush toward intolerant homogenization.

The question should not be who is guilty, as this implies that there are individuals with total autonomy over social institutions. Rather, the question should be: What are the institutional mechanisms promoting the current situation(s)? Macro structures, to a large degree, are independent of individual action. That combination of macro structures often termed the war machine by resistant culture is not in the control of a group of people, nor is it controlled by a cluster of nation-states. The very reason it is so feared is that it is out of control. It cannot be turned off, even when some of its uses to dominant culture have passed. Locating its life source is not so simple as saying that it is in the psyche of straight white men, or any other source solely constructed around the concept of agency. This is an absurd reduction that only misdirects resources toward reformist debates of minimal consequence, in that they will not change the structure or the dynamic of the war machine itself.

If the fetish for concretizing guilt and the need for genetic scapegoating can be sidestepped by leaving the bunker of imperialist ideology, it will again become possible to manufacture broad-spectrum events of disturbance. Not following the liberal code, however, does have extreme consequences. By refusing to act in accordance with the scripture of identity, one invites racist and sexist labels regardless of one's intentions. Once outside the liberal bunker, there is no security, since there are no certain enemies. No strategy can be measured in terms of probability of success. There is only speculation in this time zone, where power is liquid, with no real certainty as to which way it will flow. Consequently, acts of disturbance are gambles. The situation could possibly be made worse, much worse, by such acts, but success without the restrictions of more reforms is also possible. It is frightening to think that radical action is built upon guesswork, but if there were assurances, by what means would this work be radical? This is the much to be desired end of the heroic myths of the radical leftist as visionary or progressive thinker. All that is left is the wager, and it doesn't take genius to gamble. All that is required is the ability to live with uncertainty, and the willingness to act despite the potential for unforeseen negative consequences.

Cultural workers have recently become increasingly attracted to technology as a means to examine the symbolic order. Video, interactive computer projects, and all sorts of electronic noise have made a solid appearance in the museums and galleries, and have gained curatorial acceptance. There are electronic salons and virtual museums, and yet something is missing. It is not simply because much of the work tends to have a "gee whiz" element to it, reducing it to a product demonstration offering technology as an end in itself; nor is it because the technology is often used primarily as a design accessory to postmodern fashion, for these are uses that are to be expected when new exploitable media are identified. Rather, an absence is most acutely felt when the technology is used for an intelligent purpose. Electronic technology has not attracted resistant cultural workers to other time zones, situations, or even bunkers that yield new sets of questions, but instead has been used to express the same narratives and questions typically examined in activist art. This, of course, is not a totally negative development, as the electronic voice is potentially the most powerful in the exercise of free speech; however, it is disappointing that the technology is monopolized by interrogations of the imperialist narrative. An overwhelming amount of electronic work addresses questions of identity, environmental catastrophe, war and peace, and all the other issues generally associated with activist representation. In other words, concerns from another time zone have been successfully and practically imported into electronic media, but without addressing the questions inherent to the media itself. Again, this is a case of over-deployment and over-investment in a single spatial/temporal sector. An interrogation of technoculture itself has yet to occur, except when such investigation fits with more traditional activist narratives. As to be expected, a large amount of work is on media disinformation—the electronic invention of reality—but it is always tied to a persuasive argument about why the viewer should follow an alternative interpretation of a given "real world" phenomenon. Activists show no particular interest in questioning the cybernetics of everyday life, the phenomenology of screenal space, the construction of electronic identity, and so on.

And why should they? In the abstract sense, if power has gone nomadic, then ideology will eventually follow the same course. As speculative as it might be, with the rapid change in technology, the flowing shift of the locus of reality from simulated time/space to virtual time/space, and the undetermined speed with which this is happening, those concerned with the development of the symbolic order must ask: What are nomadic values now and what will they become? Because of cultural lag, asking questions about the fate of sedentary culture is still useful, but only if other time zones are kept in mind. Even to formulate questions relevant to electronic nomadology is difficult, since there are no theories to exploit, no histories to draw upon, and no solid issues. It is so much easier to stay in the familiar bunker, where the issues (and the parameters of their interpretation) have solidified. Here the pain of leftist authoritarianism is most intensely felt. Even though addressing issues of nomadology is clearly urgent, one fears to invoke the wrath of sedentary liberal activists by making an "insensitive" error; that fear diminishes exploration into this topic, or any other outside the traditional activist time zone. Who is willing to venture on a high-risk endeavor, knowing that the result of failure is punishment from the alleged support group?

On the practical level, this problem becomes even more complex. The hardware of everyday life cybernetics is beginning to merge, and in the most advanced time zone, that of the cyberelite, it already has. The telephone, television/video, the computer and its network structure—all these are blending into a single unit. Each of these pieces of hardware is from a different time zone, and each is thus surrounded by different sensibilities. The oldest piece is the most utopian in terms of its practical consequences in society: The telephone represents the technology closest to a decentralized open-access communication net. In the West, almost everyone knows how to use a phone and has access to one. There are even indicators that the process of decentralization that determined access to the telephone was framed as a free speech issue.* During this process, the telephone was the

---

*See Bruce Sterling, *The Hacker Crackdown*. NY: Bantam Books, 1992, pp 8-12.

best hardware for information relay available. While it clearly still had a military function, the movement to decentralize it recognized that the need for open access surpassed the need for control. It is this type of sensibility and process that must be replicated as new technologies begin to merge.

Just the opposite process occurred in the development of video/television. Although the hardware for viewing is relatively decentralized, and the hardware for production is beginning to be decentralized, the network for distribution is almost completely centralized, with little indication of change. This state of affairs must be resisted: The ideology that sanctions control of the airwaves by an elite capitalist class cannot be allowed to dominate all technology, and yet this is precisely what will happen if more cultural resources are not deployed to disturb this ideology. Cultural workers must insist on making access to electronic nets decentralized. To lose on this front is to concede to censorship in the worst way. Whether or not an artist loses h/er NEA grant because a given project is antithetical to sanctioned imperialist ideology is insignificant, compared to the consequences of merging systems of communication. This struggle will be more difficult than the opening of the phone network, since the airwaves are perceived as a means for mass persuasion. In the time of telephone decentralization, radio and film suffered defeats (access to the airwaves was perceived not as a right, but as a business), causing repercussions which are still being felt. Television took the centralized form that it did partly because of these defeats.

There is a wild card in this situation. The computer could go either way. Access to hardware, education, and networks is currently being decentralized. Unlike the telephone or television, computers have not entered the everyday life of almost every class. This primarily elite technology has sunk a deep taproot down through the bureaucratic class. The electronic service class is growing, but is far from pervasive. Hence those in lagging time zones already realize that computers are not democratic technology, nor are they considered an essential technology. This sensibility damages resistance to centralization of communication systems, since such indifference allows the capitalist elite to impose principles of self-regulation and exclusion on the technology without having to go before the public. The technology is lost before the public is even aware of its ramifications. One of the key critical functions of cultural workers is to invent aesthetic and intellectual means for communicating and distributing ideas. If the nomadic elite completely controls the lines of communication, resistant cultural workers have no voice, no function, nothing. If they are to speak at all, cultural workers must perpetuate and increase their current degree of autonomy in electronic space.

There is a more optimistic side. The computer's linkage to the telephone is much greater than to the television. In fact, the computer and telephone will probably consume cable systems. If the sensibility of decentralization can be maintained, fiber optic networks will provide the democratic electronic space that has for so long been a dream. Each home could become its own broadcasting studio. This does not mean that network broadcast will collapse, or that there will be open access to data bases; but it does mean that there could be a cost-effective method to globally distribute complex grass-roots productions and alternative information nets containing time-based images, texts, and sounds—all accessible without bureaucratic permission. It will be as easy as making a phone call.

Thus, developing systems of communication may provide another utopian opportunity. However, maintaining technological decentralization is crucial to exploiting this chance. Considering the history of utopia in ruins, the probability that this opportunity will be successfully used looks discouraging. None can predict how the technology will evolve, nor by what means the nomadic elite will defend the electronic rhizome from a slave revolt. Those engaged in electronic resistance may well be on a fool's errand, since the battle may already be lost. There are no assurances; there are no politically correct actions. Again, there is only the wager. If cynical power has withdrawn from the spectacle into the electronic net, then that is also where pockets of resistance must emerge. Although the resistant technocratic class can provide the imagination for the hardware and programming, resistant cultural workers are responsible for providing the sensibility necessary for popular support. This class must provide the imagination to intersect time zones, and to do so using whatever venues and media are available. This class must attempt to disturb the paternal spectacle of electronic centralization. We must challenge and recapture the electronic body, our electronic body! Roll the dice.

XV
1872

even when this dream reality is most intense,
we still have, glimmering through it, the
sensation that it is mere appearance

even when this virtual reality is most intense, we
still have, glimmering through it, the sensation
that it is mere appearance

XVI
1881

We operate only with things that do not exist: lines, planes, bodies, atoms, divisible time spans, divisible spaces. How should explanations be at all possible when we first turn everything into an image, our image!

We operate only with things that virtually exist: lines, planes, bodies, atoms, divisible time spans, divisible spaces. How should explanations be at all possible when we first turn everything into a virtual condition, our virtual condition!

I bought my identity, and so can you.

7

Paradoxes and Contradictions

No matter which side of the political spectrum is examined, a generalized consensus exists on the role of the individual in the formation of society, although it is phrased oppositely by each side. According to the political right, the individual must surrender h/er sovereignty to state power. From the point of view of the left, the individual must submit to enriched repression. In each case the individual loss of sovereignty is crucial. The authoritarians regard this loss as positive—the beneficent state provides the individual with security and order in exchange for h/er obedience, while radical elements see this loss as negative, since the individual is forced to live an alienating existence of fragmented consciousness. Consequently the differences between the two stem from their opposite interpretations of this act of surrender. To determine where contingent elements fall along the political continuum, one must examine the degree to which the individual is deprived of h/er personal volition and desire. Unfortunately, no presocial moment free of state power ever existed outside the imagination, so no experiential knowledge can be used to identify or to measure the qualities of liberty. For this reason, certain arbitrary assumptions must be made to fix the location of liberty anywhere on the continuum between the noble savage and the war of all against all. This either/or decision cannot be reasoned without logical error (Goedel's paradox), nor is there a history (other than *state* history) from which to make an inductive judgment. One must just decide, or act in an ad hoc or random fashion. The decision to follow any certain idea is itself a wager.

Throughout this book, the assumption is that extraction of power from the individual by the state is to be resisted. Resistance itself is the action which recovers or expands individual sovereignty, or conversely, it is those actions which weaken the state. Therefore, resistance can be viewed as a matter of degree; a total system crash is not the only option, nor may it even be a viable one. This is not to soften the argument by opening the door a crack for liberal reform, since that means relinquishing sovereignty in the name of social justice, rather than for the sake of social order. Liberal action is too often a matter of equal repression for all, in order to resist the conservative practice of repression for the marginalized and modest liberty for the privileged. Under the liberal rubric, the people united will always be defeated. The practice being advocated here is to recover what the state has taken, as well as what the reformers have so generously given (and are continuing to give).

The issue of sovereignty brings up the first contradiction to be faced here. Throughout this work, two seemingly exclusive points have been voiced: While the current situation is partly defined by information overload, it is also defined by insufficient access to information. How can it be both ways? This is a problem of absence and presence—the presence of an overload of information in the form of spectacle (presence) that steals sovereignty, and an absence of information that returns sovereignty to the individual. To be sure, information on good consumerism and government ideology is abundant. Data banks are filled with useless facts, but how can access be gained to information that directly affects everyday life? An individual's data body is completely out of h/er control. Information on spending patterns, political associations, credit histories, bank records, education, lifestyles, and so on is collected and cross-referenced by political-economic institutions, to control our own destinies, desires, and needs. This information cannot be accessed, nor can we really know which institutions have it, nor can we be sure how it is being used (although it is safe to assume that it is not for benevolent purposes). This is strategic data that must be claimed. We should be protected from the creation of electronic doubles by the right to privacy, but we are not. The right to privacy is yet another welfare state illusion in the service of the economy of desire. Specific facts about the policies and laws that promote information-gathering are not readily available, since such facts are carefully guarded by legions of bureaucrats. One needs extensive special training just to research such problems, when this knowledge could be readily available. Finally, where is the network that allows problems to be voiced on a mass scale? It does not exist.

This is a peculiar case of censorship. Rather than stopping the flow of information, far more is generated than can be digested. The strategy is to classify or privatize all information that could be used by the individual for self-empowerment, and to bury the useful information under the reams of useless junk data offered to the public. Instead of the traditional information blackout, we face an information blizzard—a whiteout. This forces the individual to depend on an authority to help prioritize the information to be selected. This is the foundation for the information catastrophe, an endless recycling of sovereignty back to the state under the pretense of informational freedom.

Dilemmas involved in the decentralization of hardware are also worth consideration. Where does Luddite technophobia stop and retrograde techno-dependence begin? This is very much a problem of finding the ever-elusive golden mean. Decentralization of the hardware invites the hazard of a techno-addiction that benefits only the merchants of technology, while centralization guarantees that electronic manipulation of individuals at both the macro and micro levels will proceed uncontested in any significant way. While the utopian claims made by the developers and distributors of new technology seem woefully transparent (after all, they are the ones who benefit the most economically), those claims are, at the same time, very seductive. The chance to be freed from the algorithms of everyday life in order to concentrate on the metaphysics of ideas is a wish worth entertaining, and has very often been vital to modern utopian theory; yet there are very discomforting elements in this vision. The economic prospects for creating such an environment are extremely bleak. If the technology were cheap enough to construct (less than labor costs), what would happen to those in the labor force? They might have plenty of free time, but no way to support themselves. To indulge the assumption that the future will be similar to the past suggests they would not fare well, since they would become an excess population. At best there would be a completely homogenized labor force, with the service sector and manufacturing sector sharing the same squalor. This scenario seems to be a return to classical Marxism in which a process of pauperization leads to two homogenized classes, with the bottom class unable to purchase the goods manufactured. The system crashes? Who can say; yet it does seem reasonable to assume that technology will not provide the utopia that corporate futurologists predict. Such predictions seem to function more in the short term, to convince people to buy technology that they do not really need, as well as to prepare future markets.

Continued reflection on the more intelligible short-term prospects of the technology of desire makes it easier to see what is immediately bothersome about technocratic promises. Take the notion of the smart house. It sounds seductive. Here is a home that runs as efficiently as its construction allows. The computer monitors household activity, and acts in accordance with these activity patterns. Energy is never wasted; it is deployed only when and where it is needed. Security systems monitor the perimeter, to alert the authorities if the property is threatened. The home is efficient and secure; it is the manifestation of bourgeois value itself. But what is surrendered when all household activities are monitored and recorded? We know that if information can enter the house, it can also leave the house, so that the price of bourgeois utopia is privacy itself. With such data available, ways for outside forces to control the household more

efficiently will also develop. Due to its surveillance components, this type of technology is another contractual trade
of sovereignty for order. What is suspect about this techno-
world is that it values consumer passivity and technological
mediation in the most totalizing sense.

This problem conjures the image of decentralization gone
awry. Decentralization does not always favor resistant action; it can have a state function. For instance, it may be
feasible for the corporate grid to provide most of the population with affordable smart machines as a marketing strategy.
The more technology available to people, and the more it
can insinuate itself into the algorithms of everyday life, the
greater the chance that it will become a market of depen-
dency. Addiction mania and hyperconsumerism are the
basis for market maintenance and expansion. The addict
always needs more. This is in part why there are such strong
punishments for addictions that do not feed corporate bank
accounts. It is intolerable to allow potential consumer
populations to focus singularly on addictions of pleasure
(food, sex, drugs). The empassioned consumer becomes
inert, rather than wandering the grid of enriched privation.
The inert consumer represents only one market of fixed
consumption—for example, a singular desire for heroin.
This kind of market is antithetical to one that remains in
flux, oscillating between accumulation and obsolescence.
The market of flux is one of entwinement—one product
inevitably leads to another, necessitating constant upgrades
and accessory purchases. One product line is interdependent with other product lines, and hence consumption and
accumulation never stop. The final goal is a diversified addiction, as opposed to one that monopolizes its consumers.

This discussion has not come full circle as it might seem at first glance. It has not gone from an apology for technology to an attack upon it. Rather, the problem being investigated is: How can technological decentralization return sovereignty to the individual rather than taking it away? Much of the answer lies in whether the technology is accepted as a means of passive consumption or as a means for active production. Passive addiction mania must be resisted; when corporate technocrats offer products or systems that seem to ride on the promises of a utopian dawn, one should scrutinize these offerings with the utmost suspicion. That which functions only "to make life easier (it all happens with the touch of a button)" is generally unnecessary. In the smart house, the computerized kitchen offers a data base on the recipes of the world. This is probably a con. Is a kitchen computer terminal really necessary? Does the service require a subscription? How often would it be used? Is it desirable to have information on daily life (cooking in this case) floating around the electronic net? Would it not be more efficient, cheaper, and private to simply purchase some cookbooks? This last question is very telling. When technology is trying to replace something that is not obsolete, one can be fairly certain that a strategy of dependence is at work. Further, continue using any technology that confounds the surveillance tactics of political economy. (In this case it is as simple as supporting book technology). Avoid using any technology that records data facts unless it is essential. For example, try not to use credit cards. An electronic record of a consumer's purchases is very precious data to the institutions of political economy. Do not let these institutions have it.

The technological artifacts and systems worthy of support are geared more toward sending out information, rather than receiving it. Desktop publishing technology is an excellent example of a system in the process of decentralization, one designed to foster active production rather than passive reception. When the technology is skewed toward reception, avoid it. (It should be noted that the strategy of entwinement is always a problem regardless of the technology chosen. Barring the total rejection of technology, the power of addiction will always be present). In the case of interactive technology, it is wise to ask, is it centralized or decentralized? If it is like the phone, and allows access to people and the information of your choice, use it—but always remember that the electronic tape could be recording. If it is centralized and spectacular, it is better to avoid it. The ability to choose an ending for a network TV show is not interaction; it is a device to keep the viewer watching. In this case, all the inventive choices have already been made. This is an example of a device designed to keep the viewer passively engaged.

To help direct technology toward increased individual autonomy, hackers ought to continue developing personal hardware and software; however, since most technology emerges from the military complex and the rest comes from the corporate world, the situation is rather bleak.

Although much of the hope for continued resistance in the techno-world rests with hackers, a contingent of resistant technocrats guided by the concerns of the radical left has yet to emerge. As mentioned in a previous chapter, this group is generally very apolitical. While they must be credited for liberating the hardware and software that represent the first moments of sovereignty in techno-culture, thereby lifting the techno-situation out of hopelessness, care must be taken not to

over-valorize them. Their motivations for producing technology oscillate between compulsion and ethical imperative.
It is a type of addiction mania that carries its own peculiar
contradictions. Since such production is extremely labor-
intensive, requiring permanent focus, a specialized fixation
emerges that is beneficial within the immediate realm of
techno-production, but is extremely questionable outside
its spatial-temporal zone. The hacker is generally obsessed
with efficiency and order. In producing decentralized technology, a fetish for the algorithmic is understandable and
even laudable; however, when it approaches a totalizing
aesthetic, it has the potential to become damaging to the
point of complicity with the state. As an aesthetic, rather
than a means of production, it can be a reflection of the
obscenity of bourgeois capitalism. Efficiency alone cannot
be the measure of value. This is one demand that the
contestational voice has been making for two centuries.
The aesthetic of efficiency is one of exclusion; it seeks to
eliminate its predecessors. Since perfect efficiency is not
attainable, and it has yet to be shown how an ascendant
system can incorporate all of the usefulness of past systems,
obscene sacrifice becomes an ever-present companion. Not
only does excess efficiency sacrifice elements of understanding and explanation, but it also subtracts from humanity
itself. Ideas, art, and passion can thrive as well, if not better,
in an environment of disorder. The aesthetics of ineffi-
ciency, of desperate gambles, of incommensurable
imaginings, of insufferable interruptions, are all a part of
individual sovereignty. These are situations in which invention occurs.

Here one stumbles upon the paradox of hacking: If hackers
must singularly commit to algorithmic thinking to be pro-

ductive, can this technocratic class be convinced to act in a manner that, at times, will be antithetical to such thinking? Perhaps the more utopian results of hacking—the decentralization of hardware and information—are in fact merely contingent elements in hacker discourse. What then is to be done? If the hackers are dissuaded from focusing on the aesthetics of efficiency, and thereby politicized, production could go down; this could in turn restrict the availability of decentralized hardware and software needed by the contestational voice. If the hackers remain focused on efficiency, that is more likely to strengthen the totalizing operations of bourgeois discourse. Treating this problem is partly a matter of redeployment. The hacker occupies a very specialized time zone, and is involved in specialized labor. Anti-company technocrats must be persuaded, by whatever available means, to enter other time zones and address the particular situations found there. Relocating hackers in other time zones should not be understood literally; instead it should lead to recombinant collaboration. That is, the characteristics of the hacker and the cultural worker should blend and thereby form a link between time zones, opening the possibilities for discourse and action across the social time continuum.

It is quite likely that decentralizing hardware (technocratic resistance) and redistributing labor (worker resistance) are not enough in themselves to intersect time zones. As already indicated, without frames of interpretation to encourage the individual's capacity for autonomous action, decentralization and redistribution could well have the opposite effect—i.e., addiction mania. The best chance to keep interpretation of cultural phenomena fluid lies in manipulating, recombining, and recontextualizing signs; when accompanied by other types of resistance, this allows the maximum degree of autonomy. Sign manipulation with the purpose of keeping the interpretive field open is the primary critical function of the cultural worker. This function separates the cultural worker from the propagandist, whose task it is to stop interpretation, and to rigidify the readings of the culture-text. The cultural worker's secondary function is to cross-fertilize separate time and/or spatial sectors, but this task has met with less success (the problem of over-deployment). The cultural worker is obligated to ferret out the signs of freedom in as many sectors as possible, and transport them by way of image/text to other locations. This transference constitutes the temporary anti-spectacle. For example, hackers have always said that the computer can grant the individual the ability to understand and to use real power. Whatever the agent commands, the computer will do. Although this may seem to be a statement of the obvious, it is questionable whether the meaning of this observation is really recognized outside the technocratic sector. If this assertion is truly understood, the possibilities for resistance dramatically increase. Populist strategies of resistance derived from reactions to the problems of early capital are only an option.

Consider the following: an activist organization decides that insurance agencies which keep records about uninsured HIV+ people contribute to discriminatory practices, and that such information-gathering must be stopped. This is not a problem of early capital imperialism, but one of late capital information codes. All the picket lines, affinity groups, and drum corps that can be mustered will have little effect in this situation. The information will not be deleted from the data banks. But to covertly spoil the information

banks, or destroy them, would have the desired effect. This
is a matter of meeting information authority with information disturbance; it is direct autonomous action, suitable to
the situation. One electronic affinity group could do instantly what the many could not over time. This is
postmodern civil disobedience: it requires democratic interpretation of a problem, but without large-scale action. In
early capital, the only power base for marginal groups was
defined by their numbers. This is no longer true. Now there
is a technological power base, and it is up to cultural and
political activists to think it through. As time fragments,
populist movements and specialized forces can work successfully in tandem. It is a matter of choosing the strategy
that best fits the situation, and of keeping the techniques of
resistance open.

Although breaks in communication lines within and between authoritarian institutions are reasonable focal points for
resistance, and it is even possible that the concrete shell of
some institutions could be completely crashed, it will still be
difficult, if not impossible, to erase all the traces of the
institution left in the rubble. Institutions, like ideas, do not
die easily. In fact, how could complex society exist without
bureaucracies? How would communication exist without
language? Irredeemable power is ongoing. Macro institutions have autonomous existence, independent of individual
action. So what is the point of resistance—why attack that
which is undefeatable? Herein lies the problem of agency.
To what degree does freedom exist for the individual? This
is a site of continuous turmoil with no satisfactory answer.
Over the past century, ideas on the degree of entrapment
have wildly proliferated. People are caught in the routinized
pathways of work, and are slaves to the demands of produc-

tion; people are caught in the iron cage of bureaucracy, and are slaves to the process of rationalization; people are caught in the domain of the code, and are slaves to the empire of signs. So much is immediately taken, from the moment the individual is thrown into the world. Even so, it is a worthy wager to assume that the individual possesses a degree of autonomy valuable enough to defend, and that it is possible to expand it. It is also reasonable to gamble that social aggregates similar in philosophical consensus can reconfigure social structures.

Of these two wagers, the former is of the most immediate concern. As the division of labor grows in complexity, individual sovereignty fades under increasing erasure, becoming a transparent transistor for social currents. Agency dwindles down to mundane choices entrapped in the economy of desire. To achieve any sense of free expression, the individual is increasingly dependent upon the latter wager. Power through numbers, though somewhat effective within the situation of early capital, is less important in late capital, as the praxis of quantity/power has hit its critical mass. Globally, an internet of unity is needed that at present is just not feasible. Even within national borders, activist organizations are encountering points of critical mass. It is a paradox; to be effective, the organization must be so large that it requires bureaucratic hierarchy. But due to its functional principle of rationalization, this rigid order cannot accommodate multiple perspectives among its members. Splintering occurs, and the organization is consumed in its own process. Perhaps it is time to reassess the idea of quantity as power. Even with the best of intentions, large groups inevitably subordinate the individual to the group, consistently running the risk of dehumanization and alien-

ation. It should now be asked, can the model used by the
nomadic elite be appropriated for the cause of resistance?

Although the nomadic elite may be a unified power, it is
more likely that this class exists as interrelated and interdependent cells powerful enough to control segments of social
organization. The interrelationship between the power cells
develops not by choice, but by nonrational process. These
cells are often in conflict, continually moving through a
process of strengthening and weakening, but the transcendental social current of late capital blindly proceeds,
untouched by the contingencies of conflict. Repression and
exploitation continue unabated. The individual agents that
labor within the cells enjoy greater autonomy (freedom
from repression) than those below them; however, they are
also caught in the social current. They do not have the
choice to stop the machinations of late capital's process.
The genetic code of these individuals is also contingent; it
is not essential to the process. They could be replaced by any
genetic sequence, and the results would remain the same,
since the power is located in the cells, not in the individual.
An individual may access power only so long as s/he resides
in the cell.

Technology is the foundation for the nomadic elite's ability
to maintain absence, acquire speed, and consolidate power
in a global system. Enough technology has fallen between
the cracks of the corporate-military hierarchy that experi-
mentation with cell structure among resistant culture can
begin. New tactics and strategies of civil disobedience are
now possible, ones that aim to disturb the virtual order,
rather than the spectacular order. With these new tactics,
many problems could be avoided that occur when resistors

use older tactics not suitable to a global context. The cell allows greater probability for establishing a nonhierachical group based on consensus. Because of its small size (arbitrarily speaking, 4-8 members), this group allows the personal voice to maintain itself. There is no splintering, only healthy debate in an environment of trust. The cell can act quickly and more often without bureaucracy. Supported by the power of technology, this action has the potential to be more disturbing and more wide-ranging than any subelectronic action. With enough of these cells acting— even if their viewpoints conflict—it may be wagered that a resistant social current will emerge . . . one that it is not easy to turn off, to find, or to monitor. In this manner, people with different points of view and different specialized skills can work in unison, without compromise and without surrender of individuality to a centralized aggregate.

*****

The rules of the game have changed. Civil disobedience is not what it used to be. Who is willing to explore the new paradigm? It is so easy to stay in the bunker of assurances. No conclusions, no certainty; only theoretical frames, performative matrices, and practical wagers. What more can be said? Roll the dice. End program. Fade out.

XVII
1890

But in this unstable, unbalanced spirit, ideas crowd on one another, and escape, and give place to others, while those that disappear still leave their shadow brooding over those that succeed.

But in this unstable, unbalanced hypertext, ideas crowd on one another, and escape, and give place to others, while those that disappear still leave their shadow brooding over those that succeed.

XVIII
1916

Animism came to primitive man naturally and as a matter of course. He knew what things were like in the world, namely just as he felt himself to be. We are thus prepared to find that primitive man transposed the structural conditions of his own mind into the external world; and we may attempt to reverse the process and put back into the human mind what animism teaches as to the nature of things.

Reality engines came to screenal man naturally and as a matter of course. He knew what things were like in the world, namely just as he felt himself to be. We are thus prepared to find that screenal man transposed the structural conditions of his own data nets into the virtual world, and we may attempt to reverse the feedback and put back into the human mind what reality engines teach as to the nature of things.

XIX
1926

Anxiety in the face of death must not be
confused with fear in the face of one's demise.
This anxiety is not an accidental or random
mood of "weakness" in some individual; but,
as a basic state-of-mind of Dasein, it amounts
to the disclosedness of the fact that Dasein
exists as thrown Being towards its end.

Anxiety in the face of cyborgs must not be
confused with fear in the face of virtual demise.
This anxiety is not an accidental or random mood
of "weakness" in some interface; but, as a basic
state-of-media of Cysein, it amounts to the
disclosedness of the fact that Cysein exists as
sliding Being towards its disappearance.

THE VIRTUAL CONDITION
Contributors:

385 B.C., Plato
60 B.C., Lucretius
A.D. 250, Plotinus
A.D. 413, Augustine
1259, Aquinas
1321, Dante
1500, Da Vinci
1641, Descartes
1667, Milton
1759, Voltaire
1776, Hume
1819, Schopenhauer
1832, De Quincey
1843, Kierkegaard
1872, Nietzsche
1881, Nietzsche
1890, Huysmans
1916, Freud
1926, Heidegger

SICK BURN CUT
DERAN LUDD

THE MADAME REALISM COMPLEX
LYNNE TILLMAN

HOW I BECAME ONE OF THE INVISIBLE
DAVID RATTRAY

THE ORIGIN OF *THE* SPECIES
BARBARA BARG

▲ **AUTONOMEDIA NEW AUTONOMY SERIES** ▲
JIM FLEMING & PETER LAMBORN WILSON, EDITORS

TAZ: THE TEMPORARY AUTONOMOUS ZONE, ONTOLOGICAL ANARCHY, POETIC TERRORISM
HAKIM BEY

THIS IS YOUR FINAL WARNING!
THOM METZGER

FRIENDLY FIRE
BOB BLACK

CALIBAN AND THE WITCHES
SILVIA FEDERICI

FIRST AND LAST EMPERORS: THE ABSOLUTE STATE & THE BODY OF THE DESPOT
KENNETH DEAN & BRIAN MASSUMI

WARCRAFT
JONATHAN LEAKE

THIS WORLD WE MUST LEAVE AND OTHER ESSAYS
JACQUES CAMATTE

SPECTACULAR TIMES
LARRY LAW

FUTURE PRIMITIVE AND OTHER ESSAYS
JOHN ZERZAN

WIGGLING WISHBONE
BART PLANTENGA

THE ELECTRONIC DISTURBANCE
CRITICAL ART ENSEMBLE

X TEXTS
DEREK PELL

WHORE CARNIVAL
SHANNON BELL, ED.

CRIMES OF CULTURES
RICHARD KOSTELANETZ

INVISIBLE GOVERNANCE: ART OF AFRICAN MICROPOLITICS
DAVID HECHT & MALIQALIM SIMONE

THE LIZARD CLUB
STEVE ABBOTT

CRACKING THE MOVEMENT: SQUATTING BEYOND THE MEDIA
FOUNDATION FOR THE ADVANCEMENT OF ILLEGAL KNOWLEDGE

CAPITAL AND COMMUNITY
JACQUES CAMATTE

 **AUTONOMEDIA BOOK SERIES**

THE DAUGHTER
ROBERTA ALLEN

FILE UNDER POPULAR: THEORETICAL & CRITICAL WRITINGS ON MUSIC
CHRIS CUTLER

MAGPIE REVERIES
JAMES KOEHNLINE

ON ANARCHY & SCHIZOANALYSIS
ROLANDO PEREZ